RESISTANCE TO THE CURRENT

Information Policy Series

Edited by Sandra Braman

The Information Policy Series publishes research on and analysis of significant problems in the field of information policy, including decisions and practices that enable or constrain information, communication, and culture irrespective of the legal siloes in which they have traditionally been located as well as state-law-society interactions. Defining information policy as all laws, regulations, and decision-making principles that affect any form of information creation, processing, flows, and use, the series includes attention to the formal decisions, decision-making processes, and entities of government; the formal and informal decisions, decision-making processes, and entities of private- and public-sector agents capable of constitutive effects on the nature of society; and the cultural habits and predispositions of governmentality that support and sustain government and governance. The parametric functions of information policy at the boundaries of social, informational, and technological systems are of global importance because they provide the context for all communications, interactions, and social processes.

A complete list of the books in the Information Policy Series appears at the back of this book.

RESISTANCE TO THE CURRENT

THE DIALECTICS OF HACKING

JOHAN SÖDERBERG AND MAXIGAS

FOREWORD BY RICHARD BARBROOK

THE MIT PRESS   CAMBRIDGE, MASSACHUSETTS   LONDON, ENGLAND

The MIT Press would like to thank the anonymous peer reviewers who provided comments on drafts of this book. The generous work of academic experts is essential for establishing the authority and quality of our publications. We acknowledge with gratitude the contributions of these otherwise uncredited readers.

This book was set in Stone Serif by Westchester Publishing Services. Printed and bound in the United States of America.

Library of Congress Cataloging-in-Publication Data

Names: Söderberg, Johan, 1976– author.
Title: Resistance to the current : the dialectics of hacking / Johan
    Söderberg and Maxigas.
Description: Cambridge, Massachusetts : The MIT Press, [2022] |
    Series: Information policy | Includes bibliographical references
    and index.
Identifiers: LCCN 2022007904 (print) | LCCN 2022007905 (ebook) |
    ISBN 9780262544566 (paperback) | ISBN 9780262372015 (epub) |
    ISBN 9780262372008 (pdf)
Subjects: LCSH: Hacking. | Computer crimes. | Capitalism.
Classification: LCC HV6773 .S638 2022 (print) | LCC HV6773 (ebook) |
    DDC 364.16/8—dc23/eng/20220711
LC record available at https://lccn.loc.gov/2022007904
LC ebook record available at https://lccn.loc.gov/2022007905

10 9 8 7 6 5 4 3 2 1

This book is dedicated to the Four Thieves Vinegar Collective

CONTENTS

SERIES EDITOR'S INTRODUCTION

Sandra Braman

Capitalism comes with big moves—accumulation, a world system, commoditization, and on. Technologies have long been fundamental to such thrusts, as has resistance to the technologies that serve capital. "Just say no," refusing or even destroying the technologies that serve capitalism, the Luddite approach, is one way of pushing back. Another is hacking—making your own technologies, using them in your own ways, and making them freely available to others. *Resistance to the Current* offers major contributions to theories of capitalism and of hacking with original conceptualizations and analyses of diverse recuperations supported by evidence from detailed analyses of quite various, little known, and quite fascinating cases.

Authors Johan Söderberg and Maxigas understand hacking to operate on three time horizons—the life cycle of individual hacker projects and their associated hacker communities, hacker projects and communities working with particular technologies in a coevolutionary dance with the industries that attempt to recuperate their creative efforts for capitalist purposes, and the epochal transformations of capitalism within which the other two time horizons unfold. "Labor" comprises three layers of workers who become involved with hacker projects outside of an employment context: commons-based peer production communities, of which hackers are only one example; the "crowds" of users of and audiences for particular technologies that appear to form voluntarily but are often and increasingly algorithmically driven; and the "clouds" of click-workers doing piece work that replicates systems created by hackers so that they can be more widely distributed. Genuine hacker autonomy requires technical expertise that is heavily dependent upon incorporation into a given collective's culture as well as, or instead of, formal education; shared values that provide the criteria upon which both ethical and aesthetic judgments can be made; and collective memory, a "usable past."

The cases examined in such detail here will be unfamiliar to most of us. Two appear on the first time horizon—Ronja was a household rooftop optical wireless router for internet access in the Czech Republic and the surrounding region and the RepRap project made affordable desktop 3D printers in the belief that this technology would end capitalism altogether by making industrialization unnecessary. Two appear on the more extended time horizon of an ecosystem of projects—hackerspaces, and the sustained use of IRC for multiple purposes across institutional environments and long-term projects despite the ultimate availability of many other technologies for the same purpose. The book as a whole operates on the third time horizon identified by these authors.

Recuperation is always a dance. The US Federal Communications Commission (FCC) repeatedly moved ham radio operators to the edge of the spectrum, knowing they would figure out how to make use of it in ways that could then be taken up industrially. For a number of years, MIT has facilitated and supported student experimentation with biotechnologies, just as it did in the 1960s with computers; the conclusion to the book—and its dedication—point the way forward, to the hacking of pharmaceuticals. The cases analyzed in depth here present a variety of types of recuperation efforts and levels of their success. With Ronja, there was in essence a standoff. RepRap was overtaken by the commercialization of 3D printing. Hackerspaces underwent multiple transformations as corporate interests sought to turn their processes and products into capital. IRC is revelatory of what can be the long-term sustenance of resistance, even inside of corporations.

The overarching past to which the authors refer includes that of engineering itself, with its iterations of progressive idealism. And of culture: it is surprising and, in fact, moving to be reminded that the Arts and Crafts movement of the early twentieth century that has so shaped still-treasured furniture, ceramics, and other household objects began with the William Morris effort to bring the production of these goods back into human hands far from industrialization for both quality of life for labor and for aesthetic reasons—an example of a hacker movement that wound up fully recuperated by capitalism. Although the focus here is economic, the highly sophisticated authors cannot stop themselves from offering cultural insights as well, noting for example changes in the use of interior lighting over the course of struggles between resistance and recuperation in hackerspaces.

We know something major has happened theoretically when what we thought was familiar becomes strange, when the unseen becomes visible, and when we understand the world in a new way. Johan Söderberg and Maxigas do that for us in *Resistance to the Current*. These are stories that need to be told.

FOREWORD

In the third decade of the twenty-first century, Johan Söderberg and Maxi-gas present *Resistance to the Current: The Dialectics of Hacking* as their elu-cidation of the digital networks that now are the essential infrastructure of our social and working lives. This wonderful combination of narrative exposition and theoretical analysis is focused upon hackers—the skilled workers of software coding and hardware making—who are the vanguard of human creativity driving forward this internet revolution. It is their experimental technologies today that anticipate tomorrow's consumer gadgets. Yet, within mainstream media and popular culture, hackers have long been endowed with a double identity of dangerous villains and rebel-lious heroes. On the one hand, they are the criminals responsible for ran-somware that threatens to shut down public utilities and for phishing emails that target people's bank accounts. Even worse, these hackers are employed by malignant foreign powers like Russia and China to steal the West's intellectual property and snoop on its military secrets. Akin to jihadi terrorists, they must be detested as wicked enemies of prosperity and security. On the other hand, hackers are also the nerds who break into computer systems to expose the hidden crimes of big business and big government. Most famously, they band together in subversive proj-ects like WikiLeaks and Anonymous that successfully mobilize public opinion against corporate plundering and imperialist wars. Like human rights activists, hackers must be admired as champions of freedom and democracy.

Söderberg and Maxigas's book counters both this demonization and romanticizing of digital iconoclasts by journalists and pundits. Instead of morality tales, *Resistance to the Current* provides a detailed examination of four projects that depict the key role of hackers in shaping the twenty-first century's network society: Ronja's community wireless networks in the Czech Republic; RepRap's self-replicating 3D printer; the social movements around hackerspaces; and the IRC messaging service. For Söderberg and Maxigas, the messy complexity of these different examples disproves the simplistic goodies-and-baddies tropes of the mass media and popular culture. More importantly, they also utilize these four historical instances to move our understanding beyond the other favored genre of hacking stories: the idealistic pioneers who either sell out or are marginalized. In a repetitive cycle, small groups of smart and cool nerds enjoy the transient exhilaration of technological invention that sooner or later is inevitably followed by—for the lucky few—the rewards of immense wealth and—for rest of them—the disillusionment of betrayed dreams. Most famously, Silicon Valley has honed its business model around this alchemic transformation of communal experiments into commercial products. The founding myth of the Californian Ideology tells how top CEOs always begin as geeky denizens of an alternative scene before appropriating its most radical idea for the dot-com start-up that makes their individual fortune. Inverting this legend, left-field thinkers mourn that the utopian possibilities of hacker groups are all too vulnerable to the competitive pressures of platform capitalism. If their invention becomes popular, they'll either be bought out by the system or get cloned by some entrepreneur. If it remains niche, they'll either burn out or go broke. Like musical subcultures, the digital underground can only keep the faith by quickly moving on to the next big thing when the last one is coopted. Hackers are techno-nomads who thrive within the Temporary Autonomous Zones outside the power structures of Empire.

As the subtitle of their book emphasizes, Söderberg and Maxigas apply dialectical reasoning to explain this repetitive narrative of invention and recuperation. Inspired by Luc Boltanski and Eve Chiapello's *The New Spirit of Capitalism*, they draw upon critical political economy to argue that the peculiarities of the hacking milieu are derived from the historically specific class position of its members. Positioned between the elite and the masses, technicians can be either labor aristocracy or petit bourgeoisie—or both at the same time. Back in the middle of the twentieth century, the development of new technologies was primarily carried out inside the research laboratories of state institutions and private corporations. In this Fordist model, scientists and engineers worked under a managerial hierarchy that rewarded dedication and loyalty with a professional income and job security. For those who wanted to succeed, their social conformity had to be confirmed by disdain for stylish clothing and bohemian attitudes. As Söderberg and Maxigas emphasize, the 1970s crisis of Fordism was the catalyst for profound changes in the socio-economic organization of technological innovation. Perfected by Silicon Valley during the 1980s and 1990s, the new paradigm dispenses with the expensive overheads of permanent staff and premises in favor of outsourcing the risk and costs of initial research and prototyping to artisanal labor and speculative financiers. Crucially, flexible markets are able to direct the self-motivated work of technicians much better than inflexible bureaucracies. Precarity and overwork are sold as individual freedom and cultural rebellion. Among the inhabitants of business accelerators who aspire to great riches, the heteronormative, crewcut, suit-and-tie look of Fordism has long disappeared. Instead, the genderfluid, dyed-hair, t-shirt-and-jeans style is now a prerequisite. Within the digital economy, everyone is—or must pretend to be—a hacker.

    Through their four examples of community wireless network, 3D printing, hackerspaces, and online messaging, Söderberg and Maxigas elucidate this dialectical relationship between autonomy and control within the twenty-first-century process of technological invention. In each instance, hackers as self-motivated workers pioneered the means and methods for the social implementation of the next iteration of the internet revolution. What distinguishes these four cases is how their participants dealt with the temptations and pressures of neoliberal recuperation. Both Ronja and RepRap thrived in their early idealistic days as small community projects, but fell apart when their handcrafted geek machine evolved into a mass-produced consumer gizmo. The squabbling and self-recrimination of their participants reveals the importance of understanding the material conditions of the class position of hackers within neoliberal capitalism.

Ironically, what felt like failure to the founders of Ronja and RepRap was successful outsourcing for their commercial successors. Subversion and disruption are excellent business strategies for making money. In contrast, both hackerspaces and IRC show how techies can preserve instances of autonomy inside the digital economy. Söderberg and Maxigas argue that the withdrawal of direct supervision over the creative process always opens up the possibility of democratically organizing collective labor. Recuperation is not inevitable, especially when commodifying digital goods or services is less efficient than sharing them. Echoing Marx's praise of workers' cooperatives in *Capital: Volume III*, the authors extol these examples of hacker self-management as premonitions of the communist future within the capitalist present. Yet, like the Old Moor, they also are well aware that these islands of socialism require a wider political and economic transformation. By contributing their specialized knowledge, hackers can help the Left to imagine how network technologies can be repurposed for the emancipation of scientific labor from the exploitative and ecocidal imperative to accumulate capital as fixed capital. For instance, in its 2017 and 2019 general election manifestos, the Labour Party's pledges to democratize the British economy took inspiration from digital workers' cooperatives and community projects both at home and abroad. Through their historical examples and theoretical analysis in *Resistance to the Current*, Söderberg and Maxigas are making their own contribution to celebrating the hacker prototype of free labor as the anticipation of social emancipation for all. Read this book!

Richard Barbrook
August 26, 2021
London, England

1

INTRODUCTION
THE DIALECTICS OF HACKING

During the pandemic, video streaming techniques turned out to be instrumental for asserting managerial authority over a dispersed workforce. In hindsight, it is hard to imagine that these techniques originated in social and technological practices of an emancipatory and critical creed. Leftwing activists put video streaming to good use in the aftermath of the Arab Spring and during the Gezi Park protests in Istanbul. A decade earlier, a company called Ustream (now IBM Cloud Video) wanted to bring state-of-the-art video streaming to consumers in a bid to disrupt the nascent social media market. The representatives of this Silicon Valley start-up sought out the cutting edge in streaming technologies. Around the turn of the millennium, they found it in the backstreets of the ancient Gothic neighborhood of Barcelona, in a hacklab called Riereta.

Riereta was known internationally for its advocacy of the Pure Data real-time audio-video synthesis system, and specifically for the development of free software libraries for video manipulation. Locally, the community is also remembered for its early morning-after parties and its polytoxicomaniac culture. The central architectural features of the space were the bar and the sound system, which were also active during the Pure Data communal patching circles that began in the afternoon and extended into the evening. The hacklab overlapped with the local squatting scene as a meeting point and sometimes as an operational base. Autonomist campaigns were broadcast on their radio show. Among many other things, Riereta pioneered promoting female participation in media production. During the last visit by one of the authors of this book, the establishment was hosting stranded Occupy activists.

As suggested by the case of Riereta, hacking is not solely or even primarily about developing new technology. At its core, hacking carries the promise of freedom. Hacking promises that freedom can be achieved by means of technical ingenuity. This ingenuity consists of the ability of hackers to repurpose tools and route around constraints and regulations. Embedded in this practice of circumvention is a critique of the predominant mode of relating to technology in modern society. Critique prompts change. Thus, it is often the case that critique serves as a motor driving technological development. And this is where the example of Ustream/IBM Cloud Video comes into our story.

Firms devise methods for harnessing the creativity of hackers. The so-called "open innovation" model anticipates the disruptive hack. Challenges to corporate prerogatives over how customers use their products are fed back into commercial product development and market research. Consequently, the computer underground as a whole is being turned into an incubator of technical and organizational innovation on behalf of the state and capital. The logic of commodity exchange, waged labor relations, and managerial hierarchies asserts itself anew over the world of hackers. The very tactic of repurposing tools has itself been repurposed. This book investigates this dialectic interplay between critique and recuperation in relation to hacker culture.

The cultural trope of the hacker as an outcast from society is increasingly at odds with the strategic position of hacking within the circuits of capital. By definition, the hacker is formally detached from contractual employment relations and professional identities. Nowadays, however, the practice of hacking is deeply embedded within occupational structures, industrial standards, and corporate innovation strategies. This follows from the centrality of the computer in the production process. Computer networks are as instrumental for managing and disciplining the workforce today as the conveyor belt and the office cubicle were in days gone by. The hacker, the emblematic outlaw of the computer age, concurrently occupies an insider position with respect to the capitalist apparatus of production.

Upon this observation hinges the wider societal relevance that we claim
for our study about hackers. The dreams and visions fermenting in the
computer underground are destined, with a lag of ten or twenty years, to
ripple through every branch of the economy. Future working conditions
and employment opportunities will be jeopardized by the technologies
hackers are tinkering with in their basements at this very moment. To substantiate this claim, it suffices to highlight a couple of well-known and
much-debated issues.

The distribution of microtasks over digital platforms creates downward
pressure on salaries in the affected sectors. Waged and regulated labor relations are dissolved and replaced with digitally mediated, piecework systems.
A pioneer in the field is Amazon Mechanical Turk (Irani 2013). The crowd-
sourcing of task allocation builds on top of methodologies first developed
in the context of collaborative, free software projects. Another example of
the same thing are integrated rating functions. Originally used in web discussion forums and content management systems, bottom-up rating functions are now deployed to monitor and incentivize employees, with the
aim of squeezing more work out of them for the same salary (Wood and
Mohanan 2019). Meanwhile, digital bulletin boards for allocating under-
used resources in local communities, often intended as an alternative to
monetary exchange (sharing rides, couch-surfing, etc.), alerted rentiers
to new and untapped sources of rent extraction. This business model was
systematized by companies such as Uber and Airbnb (Srnicek 2016). As a
consequence, the price of those resources spiked (in particular, real estate),
thus pushing out the same low-income groups that the nonmonetized bulletin boards were meant to serve in the first place. These examples suggest
that everyone stands to be affected by whatever it is that is brewing in the
computer underground. When hackers are subsumed under capital, the
outcome is an intensification of exploitative and commodified relations
throughout society.

In the same vein, hacking is increasingly becoming an instrument of geo-
political statecraft (Fish and Follis 2019). In domestic affairs, lawful intercep-
tion and instrumentalized leaks are both major issues in policy making and
parliamentary politics. While the United States and Israel use cyberweapons
such as the Stuxnet worm to sabotage Iran's nuclear program, the geopoliti-
cal tensions of a multipolar world order are projected into news reports on

Chinese and Russian sponsored hacker mercenaries. Over the last couple of decades, hacking has surged from an underground subculture to become a factor in international and domestic politics. These observations call for a reassessment of the supposed outsider position of the hacker, along with the emancipatory promise attributed to the practice of hacking.

The observation that critique is often captured and turned against its own stated goals dates from longer ago than the above examples from hacker culture. A case in point is the Arts and Crafts movement of the late nineteenth century. William Morris started it as a social movement combining a "social critique" of the appalling living conditions of the English working class with an "artistic critique" of the ugliness of industrialization. Eventually, however, the movement was sidetracked by commercial forces. It ended up as a distribution network for richly decorated handcrafts. Speaking in a different context, William Morris made a remark that could arguably be read as a self-reference to the misfortunes of the Arts and Crafts movement. The quote is taken from a short novel about a peasant uprising in fourteenth-century England. The narrator of the story (the alter ego of Morris) comments that the rights that had been fought for by the insurgent peasants long ago had now been granted to commoners in the nineteenth century. In spite of this victory, measured in the goals of ancient struggles, the living conditions of the commoner had not much improved. In some respects, they were a lot worse than during the Middle Ages: "I pondered . . . how men [sic] fight and lose the battle, and the thing they fought for comes about in spite of their defeat, and when it comes turns out not to be what they meant, and other men have to fight for what they meant under another name" (Morris 1988, 31). In this single sentence is contained a complete theory of how critique against the powers that be is recuperated and fed back into predominant structures. What we aspire to achieve in this book is to tease out the theoretical implications of Morris's quote and apply them to hackers. In the opening paragraphs, we have already hinted at one such lesson. The things that hackers are fighting for will be realized as much through their defeats as through their victories, and then this accomplishment will transpire not to have been what they meant after all.

Morris wrote the allegory of the peasant uprising as educational material for the nascent workers' movement in his own time. He sought to establish a historical continuity that brought together distinct groups under a common cause. The medieval peasantry and the industrial proletariat, although separated by a chasm of several hundred years, were fighting the same protracted battle against the same class enemy.

Following Morris's lead, we propose that hackers are continuing the struggle—under different names—for what past generations of workers once fought for. Thus, we situate the hacker movement at one end of an arc spanning hundreds of years back in time. Included under the curve of this historical trajectory are, among many others, the craft worker, the machine operator, and the white-collar computer engineer. These historical figures were battling against the same historical forces with which hackers are caught up nowadays. Now, as then, capital's domination over production is challenged on a terrain of technical know-how and access to the necessary tools.

We readily concede that such an interpretation lacks critical support in the testimonies of those who call themselves "hackers." Being a hacker is all about not being an employee. To them, emancipation means the same thing as not having any strings attached to professional and educational hierarchies and formal institutions. To be a "hacker" is the very antithesis of being a cubicle worker. In this negative way, however, the identity of the hacker is delineated by the ocean of alienated work that surrounds them on all sides.

Furthermore, even though hackers do not have formal ties to institutions, they do not escape the processes of institutionalization in a more diffuse sense. The open innovation model puts hackers to work. Firms siphon off the output from collaborative hacker projects: blue-sky thinking, incremental development work, beta testing, project management, market research, and so on. By now, the computer industry has become structurally dependent on capturing the output from hackers. This indicates an objective social relation whose validity holds irrespective of subjective testimonies of (non)class belonging stemming from self-identifying hackers.

The discrepancy between objective and subjective class positions underscores the importance of adopting a research strategy that does not take established naming practices in the computer underground as given. Indeed, this was the message of William Morris in the quote above: mistrust a name that has grown too familiar, because what it originally meant may have since changed into the very opposite! His warning resonates with the theoretical underpinnings of the concept of "recuperation." In everyday language, this word describes the recovery of a patient from illness, but it can also be used to describe the recovery of the capitalist system from periods of crisis and revolt. Coupled with the concept is a full-blown epistemology with a bearing on the naming of things. The recuperative logic of informational capitalism molds both the object of study and the analytical categories used to study that same object. Recuperation is difficult to pin down because it operates at the level of collective representations, skewing the very categories by which we try to grasp this phenomenon.

As an illustration of this rather abstract claim, let us assume for a moment that newcomers to a local makerspace are unaware that the building in question was previously referred to as a "hackerspace." Furthermore, they know nothing about the rationale behind the change of name. A year before, members of the steering committee deemed the word "hacker" to be off-putting to potential sponsors in the local business community. A few old-timers protested at the time, but they dropped out after having lost the vote. Hence, the quarrel fizzled out without leaving any traces behind. As far as the newcomers can tell, the place has never been called anything but a "makerspace." This imagined scenario, of which there are many in real life, suggests how collective memory and language are transformed under the relentless, silent pressure of recuperation.

In the above thought experiment, we allude to one episode in the evolution of hacker culture, described in more detail in chapter 5, when the concept of shared machine shops traveled to the four corners of the world. Geographical relocation brought with it a transformation in how the new spaces were conceptualized. The anarchist politics of the first wave of shared machine shops, as exemplified by Riereta, were lost in the process. This suggests that struggles over recuperation often come to a head at times of transition and generational shifts. It is thus that we interpret the tug-of-war between the Free Software Foundation and the Open Source Initiative in the early 2000s. The quarrel over naming rights and licenses was fueled by diverging opinions about the right place for ethics and politics in the advocacy for free/open software.

Just as often, however, generational shifts within hacker culture are induced by changes in industry standards. For instance, the ascendancy of social media platforms on the internet, or the diminishing prevalence of laptop computers relative to smartphones, has enticed hackers to develop new skill sets, engage with new audiences, and so on. Technology-induced transitions within the computer underground can have political ramifications as profound as those shifts that are the outcome of ideological strife among hackers.

Considering that even the historical link between "makers" and "hackers" is a fragile accomplishment, it is unsurprising that links going further back in time, such as the one connecting the "hacker" with the blue-collar "machine operator," are weaker still. It requires a theoretical reconstruction of the transformations of the labor process and working relations under capitalism for such historical continuity to emerge from the empirical material. This is not to say that we are just making things up. Our analytical procedure is warranted by the theoretical concept of recuperation. A skeptical reader might question whether we as analysts can position ourselves above the whirlwind of historical change that holds practitioners in its grip. Furthermore, how is it even possible to study a phenomenon as ephemeral as we claim recuperation to be?

To answer these objections, we look for guidance in the concept of "immanent critique," a notion closely associated with the Frankfurt School. A critique can be said to be "immanent" in relation to the justifications of the practitioners. The opposite of immanent critique is a kind of critique that passes judgment on practitioners based on criteria that are external vis-à-vis their self-understandings and value systems. External critique tends to be delivered from the elevated position of a detached observer. That being said, immanent critique diverges from the "follow the actors" approach to the study of science and technology. We do not consider the utterances of practitioners as being out of bounds for theoretical reflection. Rather, a foothold for critique is sought in the internal contradictions of practitioners' own statements and practices as these unfold over time. By making longitudinal comparisons between hacker projects, we seek to take the measure of the distance that has been traveled from one pole, the future as it was imagined in the past, to the opposite pole, what in the present is said to have been the case all along. If these poles diverge a great deal, and if that divergence is littered with accusations of betrayal and silenced voices, then we have a strong indication that recuperation is underway.

Wherever we look in the computer underground, we find strife. Conflicts abound over the specifics of license agreements, the forks of code bases, naming conventions, project leadership, and so on and so forth. Disparate as these skirmishes may seem at first glance, a closer look will reveal a highly recurring pattern. These contestations are all centered around what we elect to call the "functional autonomy" of hackers. We understand "autonomy" in the sense of collective self-determination, in contrast to the individualistic spin often put on that word. The collective in question might be a specific community of hackers collaborating on an individual development project, or it can be interpreted more loosely to refer to hacker culture in general. We take it as given that the autonomy of such collectives is never absolute. Hackers act against a backdrop of industry standards, state regulations, global supply chains, and, most decisively, the imperative of making a living by selling one's labor. No collective effort can be articulated independently of this social and material substratum.

That being said, it is warranted to distinguish between the relative degree of autonomy that a group of hackers enjoys in defining common goals and rules of conduct. In so doing, they agree on the conditions under which any member of the group, or the group as a whole, may legitimately enter into a symbiotic relationship with industry and government actors. A high degree of functional autonomy restricts the influence that external actors may exercise over the collective's decision-making process, especially when deciding the purpose to which its joint labor should be put. At stake in the many disparate skirmishes in the computer underground is the internal composition and governance structure of hackers in their coevolving relationship with institutional and industrial actors. Recuperation reduces the functional autonomy of hacker collectives, until their activity has been fully subsumed under capital's open accumulation regime. From that moment onward, collaboration among hackers will be streamlined toward the single purpose of profit maximization.

In order to speak with more precision about the social and material substratum that both conditions and enables functional autonomy, we move on to identify three pillars upon which this autonomy rests. These are: technical expertise, collective memory, and a shared value system. Firstly, having technical expertise is a prerequisite for hackers to pass independent judgments on the meaning and ramifications of any particular design choice or infrastructural configuration. Secondly, collective memory allows hackers to adopt a reflexive distance toward new technologies, as opposed to being engulfed by the presentism of the latest update. The possibility of making historical comparisons is predicated on a mythologized "usable past," to borrow an expression from Chris Kelty (2008, 64–66). The exact correspondence of the usable past with how events actually occurred is of secondary importance. What matters is that it lays down a baseline for making comparisons in the present. This baseline is drawn upon by hackers to call out deviations from the righteous path and to spur collective action. Thirdly, shared values provide the criteria for passing judgments on technology of an ethical and aesthetic kind. In order for hackers to make good on their promise to introduce something radically new into the world, it is a prerequisite that their value system must have evolved somewhat in isolation from the predominant norms and values of society at large. If not, their interventions will succumb to the same instrumental rationality that predominates in mainstream engineering culture. All three pillars of functional autonomy need to be in place for hackers to successfully detect and resist recuperation attempts.

Challenges to the functional autonomy of hackers may come in many shapes and forms. The recuperative logic of history operates at different geographical scales and at different speeds. For this reason, we propose an analytical distinction between the following three time horizons within which hacker practices and cultures can be situated and studied: 1) the full life cycle of an individual hacker project and its concurrent hacker community, 2) a landscape of hacker projects or communities coevolving together with a branch of the industry, and 3) capitalism as an evolving whole. As an illustration of the last time horizon, we can think of the passing from Fordism to post-Fordism as capital's predominant regime of accumulation. From such a macroperspective, the dialectics of critique and recuperation within hacker culture is but a minor sideshow in the ongoing tug-of-war between labor and capital. The first and second time horizons of hacker struggles unfold and acquire their true meaning within the third, overarching and epochal, time horizon of informational capitalism.

It follows from this proposition that the endgame of a particular recuperation attempt is not defined from the outset. The rules of this game change while the game is still being played. Recuperation works by surprise. Its *modus operandi* coevolves together with the boundaries of what is possible within capitalism at any given moment.

To illustrate this rather abstract claim, consider the vast amount of information stored on self-organized, underground web forums. Those who once posted the comments, photos, trip reports, and so on had no idea about the utility that third-party actors would eventually derive from the information. Only much later, with the advancement of techniques for mining heterogeneous data sets and cross-referencing them with other kinds of information, geolocations, biometric markers, and so on, have those postings been turned into monetizable assets.

This observation alerts us to the need for anticipation. It warrants the analytical distinctions we are proposing between different time horizons in the study of hacking. In order for us to anticipate future recuperation attempts, hacker projects must be investigated within longer time series than the single, one-off case study.

It goes without saying that it is we as scholars who are postulating these analytical distinctions. Where the line is to be drawn between one time horizon and another can only be decided on a case-by-case basis. It would run counter to the grist of our argument if we were now to pretend that we had pinned down recuperation in a bullet point list. We would then be in for an unpleasant surprise. On the assumption that recuperation works by surprise, its anticipation takes something akin to Immanuel Kant's definition of an aesthetic, reflexive judgment. What makes a work of art beautiful, according to the philosopher from Königsberg, cannot be explained by means of any list of pregiven concepts. It takes experience and practice to acquire the knack of passing a timely and accurate aesthetic judgment. In the context of hackers, Kant's third critique can be translated as follows: we have to train our noses to "sniff out" recuperation attempts. One of the purposes of this book is to contribute to this learning process among hackers.

The merit of adopting this theoretical lens hinges on that it brings more explanatory clarity to the empirical study of hackers than it obfuscates. In accordance with this methodological precept, we will put our theoretical framework—which is set out in more detail in the next chapter—to the test in four historical case studies. The four cases have been selected with an eye to illustrating different possible outcomes of resistance to recuperation attempts. Furthermore, the four cases are set within different time horizons within which recuperation unfolds.

The first empirical chapter plays out within the time horizon of the life cycle of a single hacker project and community. It investigates the Ronja project, where a free-space optical (FSO) device was developed to connect computers using visible light. It flourished in the Czech Republic and in neighboring central and eastern European countries a few years before the notion of "open hardware" had been coined internationally (Söderberg 2010). While Ronja outperformed competing technologies at the time, internal strife over the direction of the project and the design of the device blocked further progress. The community dissolved as would-be entrepreneurs channeled their efforts into private development forks. This recuperation attempt failed insofar as the demise of the Ronja project did not result in any commercially viable product.

The second case study also takes place within the time horizon of an individual project, but this time it showcases a partially successful recuperation attempt. At its inception, the open-source 3D printer called RepRap was hailed as a machine to end capitalism. The spread of a ubiquitous manufacturing unit would render markets in most consumer goods superfluous, as the items could be printed at the cost of the materials (Söderberg 2013, 2014). The project was first hosted in university departments, then in hackerspaces. Soon, start-up firms began to compete with their branded derivatives of the 3D printer (Bits from Bytes, MakerBot), and, in no time, major industrial players (Stratasys, 3D Systems) asserted control over the niche market for desktop 3D printers. This reveals a successful recuperation attempt, insofar as a product innovation and a corresponding consumer market did emerge from the collaborative hacker project. In spite of this, we consider the recuperation process to have been incomplete. A potentially more disruptive organizational innovation by the RepRap community—the possibility of distributing physical manufacturing to a network—could not be assimilated by the industry. However, this was not due to active resistance from the hacker community. Rather, its failure was due to lacking the means for enforcing quality control and worker discipline at a distance.

The third case study unfolds within the second, more extended time horizon involving a landscape of projects and communities—corresponding to a whole branch of industry actors and innovation structures. This is the story of hacklabs becoming hackerspaces, hackerspaces inspiring Tech Shops, and Tech Shops evolving into start-up incubators and accelerators (Maxigas 2012, 2015). While hackerspaces are membership-run clubs for technology enthusiasts, incubators and accelerators use the culture of hackerspaces and their hackathons to jump-start companies and/or attract regional funding from governments. This case serves to demonstrate the real subsumption of some hacker practices that extends beyond any individual project or product. Rather, it signifies the reorganization of the labor processes of hackers.

The fourth case also runs its course within the second time horizon, involving a landscape of projects. This time, however, recuperation processes were successfully resisted. The case study is about Internet Relay Chat (IRC), the backstage communication infrastructure of peer production communities. This chat infrastructure was set up in 1988 and survived the commercialization of many forms of community-run media during the dot-com boom. It emerged into the new millennium as a growing community that has accommodated the needs of free software developers, Wikipedia editors, hackerspace members, and Anonymous hacktivists during the last decade (Maxigas 2017; Latzko-Toth and Maxigas 2019). In the meantime, generations of chat technologies came and went, while corporations such as Microsoft tried to capitalize on the success of IRC, but to no avail. Therefore, the IRC infrastructure and its community offers a hopeful spark, suggesting that the recuperative logic of capital can be successfully resisted.

To summarize, the first case study ends in a stalemate and the dissolution of the project; the second one showcases how failure to resist a recuperation attempt is vindicated in a marketable innovation; the third culminates in the subsumption of a whole segment of hacker culture into capital; and the fourth demonstrates hackers resisting a series of recuperation attempts. Rather than trying to exhaust all the possible outcomes, however, we hope that the case descriptions will convey the contingent elements of history.

There is a reason why we have not dedicated an empirical chapter to the third time horizon in our analytical scheme, capitalism as an evolving whole. The totality of social relations does not offer itself as an object for case study types of inquiry. To address capitalism directly, we would have had to write a different kind of book. That being said, the four studies add up to tell a tale about capitalism. Our cases are all located at the far end of a historical arc, which began with the golden age of hacking in the 1970s and 1980s. Back in those days, confidence was running high in the computer underground that state and capital could be outwitted on the leveled playing field called "cyberspace." Our small sample of case studies captures the afterglow of that collective self-confidence and utopianism. Equally so, the case descriptions bespeak the closure of this historical window. Above and beyond the individual gains and losses in the innumerable struggles that have come to pass, something more fundamental is brought into question when hacking is situated within the third time horizon. At stake is the very essence of hacking, the promise that freedom can be derived from the practice of repurposing tools and routing around constraints.

What remains of this promise after the practice of hacking has been turned into a motor to drive capital accumulation? Hacking offers a textbook example of how critique and resistance furnish capitalism with the means for overcoming its own limitations and blockages. Saying this is not to throw one's hands in the air. The upside of adopting a *longue durée* perspective on hacking is that it concurrently implies that the figure of the hacker is but one instantiation in a much longer series of struggles. This encourages us to expand the search outward from hacker culture to look for allies in the many other places that have been caught up by the same historical forces. Ending on this note, the concluding section of the book scouts out the upcoming sites of contestation that await us in post-pandemic times: the extension of hacker practices to the manufacturing of pharmaceuticals.

2

THEORIZING CRITIQUE
AND RECUPERATION

Starting from the observation that "hacking has been hacked," in the following pages we develop an interpretative framework for studying the dialectical reversal of hacking. The core of our theoretical argument is borrowed from Luc Boltanski and Eve Chiapello's book *The New Spirit of Capitalism*. With them, we take as our point of departure the idea that capitalism advances by incorporating the critiques directed against it. Critique, once recuperated, is transformed into a source of innovation and legitimacy for capitalism (Boltanski and Chiapello 2005). It goes without saying that this is a two-way process. At the same time as criticism (and the critics) are incorporated into dominant structures, those structures are also transformed by the critique. A dialectic between critique and recuperation unfolds.

Boltanski and Chiapello developed a theory about the general, historical logic of capitalism. The dynamic that they describe in general and abstract terms is apparent in the empirical field of hacking. Silicon Valley owes its existence to a computer underground that was spawned from the American counterculture of the 1960s (Barbrook and Cameron 1996). A number of historians of technology have filled in the details about how cyberculture and counterculture merged and furnished the nascent computer industry with communication standards, working practices, consumer preferences, and so on (Zandbergen 2011; Turner 2006; Liu 2004). Contemporary examples of the same kind of symbiotic relationship between the computer underground and various branches of the computer industry are easy to come by, as we will discuss in more depth in the empirical chapters (see also Delfanti and Söderberg 2018).

In addition to the work of Boltanski and Chiapello, our interpretative framework also draws upon a number of supplementary theoretical resources. Primary among these are social movement studies, autonomist Marxism, and labor process theory. Social movement studies, especially where it intersects with Science and Technology Studies (STS), offers a rich source for reflecting upon how activist groups achieve their demands at the price of becoming part of the institutional arrangement that they railed against. The contribution from autonomist Marxism is twofold. Firstly, it has advanced an interpretation of technological change as the outcome of capitalist restructuring in response to class struggle (Dyer-Witheford 2015). Secondly, it has theorized the "social factory" (Tronti 1979) or the "factory without walls" (Negri 1989). The basic idea is that the factory—and with it, the contractual employment relation—no longer delimits the site of capitalist value production. Scholars in new media studies are putting empirical flesh on these theoretical bones by describing how media companies have become structurally dependent on exploiting the free labor of fans, audiences, users, and so on (Gill and Pratt 2008; Fuchs 2014). Hackers fit neatly into this same line of argument. In the tradition of labor process theory, finally, we find empirically grounded insights into how management and workers wrestle over who is in control of the workplace and the paramount role of skills and technological design in determining the outcome of those struggles (Böhm and Land 2012; Söderberg 2019).

Based on a synthesis of these theoretical resources, we propose an interpretative framework for conducting fine-grained, empirical studies on the dynamic of critique and recuperation within hacker culture. Hacker projects brim over with tension and strife: alleged violations of licensing terms, bickering over the correct names to call things, endless negotiations over what software tools are proper, and so on. These skirmishes are not isolated, random events. They make up a larger pattern of struggle over the conditions under which hacker culture may enter into relations of mutual dependency with industry and government agencies. Through those struggles is determined the relative degree of functional autonomy of individual hacker projects—and of hacker culture at large—vis-à-vis an exteriority.

By the word "autonomy," following an established tradition in political philosophy, we mean the capacity of a collective to determine its own goals and rules of conduct and then to remain true to those goals and rules no matter what the cost. A group of hackers enjoys a high degree of autonomy when they can determine the methods and purposes of their collaborative endeavors and future collective existence. The idea of autonomy is not falsified by the observation that hackers derive a livelihood and gain political leverage from entering into symbiotic relationships with industry, government, and academia. With our interpretative framework, however, attention is directed toward the risks and benefits of such alliances for the autonomy of the collective.

When autonomy is weakened, hackers can more readily be subsumed under an "open innovation" model. They are thus turned into a steady source of blue-sky ideas and problem-solving work. In saying this, we intend to turn the tables on the vast literature about "open" or "user" innovation (Chesbrough 2003; von Hippel 2005). In our reading of the situation, innovation is what happens when hackers fail to resist recuperation attempts. Indeed, the open model for procuring innovations from hackers (as well as users, fans, etc.) is the spirit of contemporary capitalism, to reconnect to Boltanski and Chiapello's terminology: it is how production is organized outside the factory walls.

In order to discuss the recuperation of critique with greater precision, we propose a number of analytical distinctions in this chapter. Firstly, we introduce three time horizons, or temporalities, within which the evolving relationship of mutual dependency between hackers and industry may be studied. Secondly, we survey the factors required for putting up an effective resistance to recuperation within the recuperative logic of history, which we elect to call the "three pillars of autonomy." These pillars are: technical skills, historical memory, and shared values. Thirdly, we propose a typology to talk about a budding social division of labor outside of formal employment relations. These are: communities of peer producers (such as hackers), crowds of users and audiences, and clouds of click workers. The refinement of this division of labor is one of the things to which recuperated hacker projects are contributing: for instance, by developing the tools used by management to oversee a decentralized workforce. "Commodification" refers to the process of enclosing commonly held resources and inserting them into market circulation. The antonym of commodification is "communization." Resources nominally owned by the employer are rerouted by the employees to serve as common infrastructure and to support the reproduction of a shared culture.

These conceptual tools cannot stand on their own. Their analytical worth must be proven through empirical case studies, such as the ones we expound in the following chapters, in which we present four historical cases in depth. The point of this exercise is to alert us to likely developments ahead, and so we conclude the book with an anticipation of the upcoming circuit of struggle.

WHO COUNTS AS A "HACKER"?

Before we can start the theoretical discussion proper, we need to clarify the words and definitions that we will be drawing upon, starting with the key word, "hacker." Different uses of this word are circulating in the academic literature, often bearing witness to the disciplinary fields from which those studies derive. The different strategies for defining hackers can be grouped into four general types. In empirically oriented fields, scholars typically allow the people calling themselves "hackers" to define the word. If a scholar creates a definition based on some theoretical and disciplinary background knowledge, then hacking tends to be assigned a meaning stemming from studies of youth subcultures, social movements, or class analysis. We discuss the strengths and weaknesses of each approach in turn.

Definitions based on self-appointed hackers usually take their basis in an excerpt from the Jargon File, a widely recognized lexicon of hacker slang. The first entry for "hacker" reads: "A person who enjoys exploring the details of programmable systems and how to stretch their capabilities, as opposed to most users, who prefer to learn only the minimum necessary" (Steele and Raymond 1996, "hacker"). Three more entries follow, stressing the hacker's aptitude for programming. In addition, some general characteristics expected of an individual claiming to be a hacker are described, such as enthusiasm, curiosity, and the like. This offers a point of departure for studies of hackers, but it does not allow the scholar to advance very far beyond what the practitioners already profess to know. Theoretical shallowness is a common weakness in many, otherwise empirically solid, exposés of hackers (Benkler 2006; Moody 2001; Weber 2004). This approach is problematic when the self-representations of the hackers are so faithfully conveyed by the academics that the practitioners' exclusions and omissions are reproduced and stamped with scholarly insignia. A case in point is the dismissive description of the "cracker" in the Jargon File, as someone who "breaks security on a system" (Steele and Raymond 1996, "cracker"). While free software development is associated with positive values, such as information sharing and transparency, this does not tell the whole story about a subculture that is just as much about secrecy and stealth. The latter aspects are strategically left out in the accounts given by many hackers, who are preoccupied with improving their tarnished public image.

A definition of hackers that is more loosely tied to their self-representations can be found in the field of cultural studies. Here, hackers are interpreted as one youth subculture among many. This perspective, put forward by Douglas Thomas (2002), has a lot to offer to our discussion. After all, subcultures are all about defining who belongs to the group and who does not. Of particular interest for our purposes is the fact that Thomas foregrounds how the hacker milieu differs from most other subcultures. The identity of a hacker is bound up with a practice rather than a style. He argues that this endows hackers with a greater amount of self-determination in relation to external influences than style-based subcultures, which are more easily swayed by commercial forces. In line with the cultural studies tradition, however, he understands this in terms of a generational struggle (Thomas 2002). The blind spot of the cultural studies field is political economy. Thomas's definition fails to take the measure of the successive integration of hackers into professional life as they come of age. Such a definition excludes the main topic of our investigation from the start, which is how hacker culture negotiates the perils and opportunities of subsumption under the computer industry's schemes of value production (Lakhani and Wolf 2005).

Social movement studies is more attentive to the political stakes involved in hacking. Another advantage of this approach—of which Paul Taylor and Tim Jordan are the chief proponents—is that it aims to describe how hackers constitute themselves as politically conscious subjects capable of collective action. This outlook makes a lot of sense when studying how hackers mobilize against intellectual property laws, state surveillance, and the like. The approach is fitting for studies of WikiLeaks, Anonymous, and similar political groups with strong ties to hackers and a strong support base in hacker culture more broadly. There are many other aspects of hackerdom, however, that will be missed if it is studied using an interpretative grid borrowed from social movement studies. The stereotypical hacker does not perceive him- or herself as a political activist and is more likely to vehemently deny such comparisons. This is a problem for social movement theory, which stresses raised consciousness and articulation by its actors. One risks losing sight of what makes this setting special, what Gabriela Coleman once named the "political agnosticism" of hackers (Coleman 2004; Coleman 2012, 187–189). If pride of place is given to a relatively small group of activists who fit the bill, to the detriment of the rank-and-file, free software programmer, something important has been missed.

A fourth approach to the study of hackers draws on the tradition of political economy and class analysis. One common variant that we want to avoid is to declare that a segment of the population constitutes a new class of hackers in its own right, on a par with the working class (Wark 2004; for a critique, see Barbrook 2006). Such claims, made in disregard of what self-described hackers say or do, bring no analytical clarity to the discussion. More promising in our opinion is a class analysis that starts out from labor process theory and empirically oriented workplace studies. In the tradition of labor process theory, starting with Harry Braverman's classic *Labour and Monopoly Capital*, there was always an astute awareness of the need to keep the definition of the working class as dynamic as the everchanging labor process. This warrants our analytical procedure of including free software developers, hackers, and related kinds of hobbyists as a subset within the occupational structure of waged programmers and engineers. We acknowledge that the status of being a nonemployee is foundational to their identities, and so our categorization is gainsaid by subjective testimonies. That being said, empirical support is readily found in a computer industry that has grown to be structurally dependent upon extracting value from their activities (Kirkpatrick 2018; Liu 2004). The large discrepancy between, on the one hand, hackers' identity of being nonworkers and, on the other, the objective role that they fulfill within capitalist circuits of accumulation is what warrants our inquiry into the dynamics of critique and recuperation. Thus, we offer an interpretative framework for empirical studies of the processes whereby hackers are turned into sources of revenue and innovation for the industry, or resist such attempts.

Then, who are we talking about when we discuss "hackers"? The answer to this question is further complicated by the fact that the phenomenon of hacking is as fluid as the technology to which it relates. Hence, any definition of the hacker must not be too closely linked to the use of any single device or infrastructure. In the early days of hacker culture, tinkering with hardware and interfering with telephone networks were central features (Levy 1984; Lapsley and Wozniak 2013). In the 1990s and early 2000s, to do hacking became synonymous with writing software code and meddling with network protocols. The practice was often inscribed in the then very popular master narrative about a "coming of the information/network society" (following Castells 1996). Subsequently, the confinement of hacking to a realm of "bits" was positively affirmed and contrasted against the old, industrial "world of atoms." This dichotomy is not salient today. A definition of hacking that excludes tinkering with hardware and interventions in biology would miss out on much of what is going on in hacker spaces today (Seravalli 2012; Kostakis, Niaros, and Giotitsas 2015; Delfanti 2013).

Conversely, however, the specificity of the hacker vis-à-vis other subcultures and movements would be lost if all references to technical practices were abandoned. An example of this is when artists involved in what was once called "culture jamming" claim to be doing "semiotic hacking." We therefore insist on the connection to technical practices. Without it, we cannot make sense of the meritocratic values within hacker culture. Being skilled is key when hackers distinguish themselves from ordinary computer users, variously labeled as lamers, n00bs, AOLers, and so on. Furthermore, hacking cannot be extended to include just any technology whatsoever, as this would make the word "hacker" synonymous with "tinkerer" and "inventor" or "hobbyist." There must be a connection on the symbolic plane, however remote, with practices relating to information processing and information security.

Open hardware development qualifies, for instance, because it borrows its methodologies, licenses, values, and cultural tropes from the established tradition of free software development and computer hacking (Ackermann 2009; Powell 2012; Söderberg and Daoud 2012; Lindtner, Hertz, and Dourish 2014). Continuing along the same lines, the main argument of Delfanti's monograph on biohackers is exactly that "[o]pen biology is embracing values and practices taken from the world of hacking and free software [so that] science is experiencing the same type of differentiation and complexity shown by hacker cultures" (2013, 12). We concur with his observation, that with each subsequent wave of objects to which hacking is applied, the milieu seems to be becoming ever more integrated into the capitalist system. Hardware development is ultimately reliant upon a global supply chain, which might explain why hacker culture has come to hybridize with the shanzhai innovation ecosystem—likewise with DIY biology, which has developed in tandem with spinoff companies eyeing up venture capital and seeking certification from the authorities (Keane and Zhao 2012; Tocchetti and Aguiton 2015). The title of an article written by Delfanti shortly after the first dedicated biohacking conferences took place and the first community spaces for biohackers became established puts the question upfront: "Is do-it-yourself biology being co-opted by institutions?" (2014). Each time there has been a shift in the center of gravity within hacker culture, from phone phreaking and hacking to free software development, and from software to hardware and biology, a shared body of cultural tropes and values has been passed on, although, we venture to say, in an increasingly attenuated form. Thus, it is possible to grasp the elusive subject of hacker culture as a historical formation rather than as a purely theoretically motivated conception. To summarize, we base our analytical distinctions on the genealogy of collective representations among the people who identify with the "hacker" moniker.

This pronouncement begs the question: On what analytical level do we intend our distinctions to be applicable? We are obviously not content with simply referring to the individuals who call themselves "hackers." We also analyze the distinct historical trajectories and group dynamics of various separate collectives. Our musings about functional autonomy hinge on how the collective entity has been defined in the first place. We make use of a variety of designations: "hacker culture," "free software movement" and "open hardware movement," "hacker projects" and "hacker communities," and "computer underground" and "geek public." The different names carry different connotations and bring into play differ-ent analytical levels. We draw upon different terms depending on what aspect of hacking we want to emphasize at particular junctures in our argument. For instance, when we discuss a common identity that encom-passes the full spectrum of phenomena associated with hackers, we use the phrase "hacker culture." Under its umbrella gather past and present instances of hacking, including phone phreaking, the warez scene, free software and open hardware development, hackerspaces, DIY bio, and so on. When the emphasis of the argument is instead on the contradistinc-tion between them and the surrounding society, we refer to the "computer underground" or "hacker subculture." The variation in language use is motivated, as we argued in the paragraphs above, by the need to explain hacking from a combination of disciplinary and theoretical perspectives.

Our use of the terms "hacker project" and "hacker community" war-rants some additional comments. When saying "hacker project," we are referring to a delimited subgroup within the larger hacker culture who are dedicated to the development of a single software or hardware artifact. A hacker project is closely associated with a recognized project leader, com-mon engineering goals that have been defined in advance, a single license under which the output is published, a dedicated discussion forum or a cluster of such forums and websites, and a core set of developers and beta testers. When we want to bring the developers and users into focus, rather than the artifact and the development process, we say "hacker commu-nity." The word "community" is loaded with many associations and con-ceptual ambiguities. For lack of a better word, we use it to name the loose constellation of developers and users who frequent a discussion forum, know each other from before, and jointly care about the future of the development project in question.

Occasionally, we compare communities of hackers with gatherings of amateurs, fans, hobbyists, and so on. The latter coalesce around similar kinds of generative practices as hackers, but without identifying them-selves with the overall hacker culture. The emergence of such practices in many different sectors of society has been intensely debated under the heading "commons-based peer production" (Benkler 2006, chap. 3; Rigi 2012). Keeping to the established terminology in the literature, we refer to a peer production community as a generic, catch-all term that is not beholden to the subjective identity constructions of the practitioners. As we move on in this chapter to unpack our theoretical arguments, however, we will propose a tripartite typology for making distinctions within commons-based peer production. This typology is: "communities of peer producers," "crowds of users/audiences," and "clouds of click workers." All of these categories designate abodes of production located outside of the formal employment contract and the factory gates. For similar typologies inspired by different theoretical traditions, see Jenkins, Ford, and Green (2013) or Shirky (2008). In keeping with the overall, interpretative framework of this book, we argue that meaningful distinctions can be made in the degree of autonomy (or lack thereof) that each group formation exercises over its own collective being in relation to external interests (i.e., state and capital). The classification of communities, crowds, and clouds as different constellations of value-producing activities corresponds to the emerging trend toward a division of labor outside of the wage labor relation.

Finally, with the term "hacker movement," we indicate a broader sweep of hacker projects and hacker communities that are bound together by common goals and formal support structures. This claim can be illustrated with the free software movement. A wide range of hacker projects, dedicated to the development of a single free software application or operating system distribution (such as, for instance, the Debian project), unite under this movement. They share the same software tools, content management systems, and licenses, go to the same conferences, and so on. This shared technical and organizational infrastructure is duplicated in shared concerns and discussion topics. The free software movement is broader in scope than any individual, free software development project, while at the same time being more restricted and bounded than the reference to a "hacker culture," as the latter must also include open hardware, DIY biology, and so on.

As a final remark, we note that our conceptualization of hacking as a collective entity, operative at different analytical levels, comes close to the approach of Christopher Kelty. Acknowledging the heterogeneity of the phenomenon, he also makes a compelling case that scholars need to refer to the commonality of attributes that fall under the umbrella of "hacking." He then draws a parallel with the eighteenth-century notion of the "public." Just as the public in those days defined itself in opposition to absolutism, the notion of being a counterweight to the powers that be looms large in the self-presentations and identity constructions of hackers. This is a crucial point, because, although the concept of a public is sufficiently vague to include a range of phenomena, it is coherent enough to allow for collective action. The common identity of hackers across innumerable variations and differences is verified by the fact that, from time to time, they can come together and act in concert. Kelty's reliance on collective action as the defining trait of hackers resonates with the approach that we are adopting in this book.

From Kelty's work, we have also borrowed the idea that the public acts "recursively" to create the material conditions upon which its own, continued existence depends. The eighteenth-century public emerged in close association with the technologies of its time, chiefly coffee houses, book printing, and newspapers. Likewise, hackers rely on free software, transparent and compatible network standards, open hardware, and so on. In contesting intellectual property rights and promoting open infrastructures, hackers are not solely striving to realize their vision of a good society. They are concurrently safeguarding the legal and technical conditions for their own continued existence as a collective of peer producers. This fits neatly with our primary concern in this book, to describe how hackers resist recuperation attempts and maintain a relative degree of autonomy vis-à-vis an exteriority upon which they are nevertheless dependent.

THREE TIME HORIZONS OF CRITIQUE AND RECUPERATION

The coupling of "critique and recuperation" describes a dynamic wave of historical change propelled by struggle. Critique can take many different forms, but in the context of this book we are chiefly referring to critical engineering practices (Oliver, Savičić, and Vasiliev 2011). Through such practices, dreams about a radically different society become embedded in alternative product designs and diverging pathways in the development of technology. Critique, thus understood, spurs capitalism to evolve in new directions. By transforming itself, capitalism absorbs the disturbance and turns it into a material and organizational infrastructure for the continued accumulation of capital. So that we may talk about this dynamic with more precision, we proceed to distinguish three time horizons within which recuperation processes unfold.

The shortest time horizon encompasses the life cycle of an individual technology development project and its associated community of developers. In the case of open-source desktop 3D printers (discussed in chapter 4), for instance, the starting point was the launch of the RepRap project in 2004. This project drew to a close in the early 2010s, when leading corporate players consolidated their control over the niche consumer market for desktop 3D printers. The vision that underpinned the development project in the early days, concisely captured in the project's byline, "wealth without money," had conclusively been recuperated when derivatives of the RepRap 3D printer started to circulate as commodities.

The second time horizon spans a landscape of hacker projects and communities evolving in concert with an associated branch of the industry. A case in point is the duration of the free software movement, from its inception in the mid-1980s until today. Although the free software movement has not come to an end, it can be meaningfully delimited from its successor, a social movement dedicated to the development of open hardware. Whereas the free software movement is closely associated with the computer industry, the open hardware movement is evolving in tandem with products and industry standards stemming from the consumer electronics industry. Recuperation processes operating within this time horizon shape the framing conditions for individual hacker projects. They tilt the balance of forces in upcoming, local struggles against recuperation attempts.

The third time horizon relates the phenomenon of hacking to epochal transitions in capitalism as an evolving whole. Hacker culture is at one and the same time a historical product of the capitalist economy and a—tiny but strategically placed—contributor to the further development of this economic system. We speak at this level of abstraction when, for example, we relate the birth of the computer underground in the 1960s and 1970s to the transition within capitalism from Fordism to post-Fordism.

These analytical distinctions are needed because the dynamic of critique and recuperation plays out simultaneously across many different geographical scales and time horizons. Recuperation attempts within a local hacker project advance in a piecemeal, iterative fashion. The microstruggles over an alleged license violation or an ideologically informed design choice, for example, add up to shape capitalism as a whole. Short-term dynamics of action and change acquire their meaning within a wider frame of reference corresponding to the succession of capital's accumulation regimes. Recuperation attempts can be meaningfully resisted in a local setting—as, for instance, when hackers refuse to adopt one or another commodified and black-boxed industry standard. The catch is, however, that the significance of this refusal is conditioned by forces that surpass the local setting. Only with reference to the grander scale of things can judgments be made about whether or not to use a particular consumer product or choose one development fork over another.

Furthermore, recuperation does not exclusively impact upon the subjugated hacker project. Once a project has been recuperated, it furnishes capital with product innovations and ideas for organizational reforms that can then be exported to other sectors of the economy. Given the global and social division of labor, it is only to be expected that the most coercive side of recuperation is not experienced by those who are most directly concerned with it (i.e., the subjugated hacker community). The lowest tier of the labor market bears the brunt of recuperated hacker projects. Two examples will suffice to make the point: the digital platforms for dividing up and distributing piece rate work in the "cloud" and the new methods for surveilling and disciplining workers based on digital rating systems. In order for us to catch a glimpse of the geographically dispersed and protracted fallout from recuperation, we must not restrict our investigation to one-off case studies of technology projects. Hacking needs to be studied within an interpretative framework that can be scaled up to encompass capitalism as an evolving whole.

By historicizing the framing conditions of individual hacker projects, we arrive at the conclusion that the outsider position to which hackers often lay claim is always already inside the larger whole of historically developed, capitalist relations. Putting it differently, hacking is (partially) recuperated from the outset. The notion of a free-floating subject position located "outside" the social totality of capital is illusory. Connected to this ideological notion of the outsider is another, equally problematic, idea. Namely, that a vortex of disruptive innovations will flow from the hacker to uproot the constituted order and tear down incumbent interests. We contend that this Schumpeterian fantasy works like a trap for capturing value. The disruptiveness of hacking has been anticipated by the open innovation model.

Something else that is ruled out by our historical approach is the notion of a pristine, golden age of hacking that at some point was forsaken. We endorse the commonplace notion that the past is an ex-post construction. Even so, we cannot choose to not posit a point of origin that has culminated in our present mode of existence. We need "usable pasts" as a baseline for making comparisons with the present, to pass normative judgments and take directions toward a more desirable future (Kelty 2008, 64–66). We reject the shallow wisdom of nominalism—whether it comes in the form of commonsense empiricism or as convoluted poststructuralism—and the unqualified celebration of contingency that is common to both. Often this outlook is an overreaction by those who previously longed for an absolute foundation, only to become disappointed and sink into despair. The right lesson to be drawn from the argument about contingency is that, although history is all that we have to hold onto, this is sufficient for the purpose of guiding collective action. In saying this, we endorse a key insight in Hegelian philosophy and in its modern-day heir, the immanent critique tradition (Antonio 1981). Although we do not dwell on these references in what follows, for the record, we declare this to be the philosophical backbone of our observations about the dynamic between critique and recuperation.

FIRST TIME HORIZON: RECUPERATION OF
A SINGLE HACKER PROJECT/COMMUNITY

The first time horizon within which we can observe the dynamics of critique and recuperation in hacker culture is that of the life cycle of an individual development project and its concomitant community of developers. Recuperation attempts operating within this time horizon strive to enclose product innovations and other tangible goods stemming from the common development process. This corresponds to the "enclosure of the commons" idea, which is widely diffused among practitioners. One-off enclosures add up and work together as step-by-step advancements by the industry to subsume the hacker project in question under an open innovation model. Thus, hackers are turned into a steadier and less risk-inducing supply of disruptive innovations and piecemeal development work. When we investigate hacking at this analytical level, we conduct case studies of the life cycle of a single technology. The temporal aspect is key, because it is in the displacements that occur from start to finish in the testimonies coming from the project's participants that we detect signs of a recuperation process.

The last point accords with the insight in STS that an adequate study of a technology should pay as much attention to failures and dead ends as to successes. It will then transpire that the stated goals of a project, the individual motives for being involved, and the design itself, have all metamorphosed during the course of the project's life cycle (Edgerton 2008; van Oost, Verhaegh, and Oudshoorn 2009). This point is warranted by the many success stories about start-up companies and innovations that fill page upon page of airport and management literature, as well as being the default narrative in some academic fields, such as innovation studies. These success stories are narrated with the commercial breakthrough as their referential cutoff point. What happens to the community of developers and users who incubated the innovation in the first place becomes unimportant after the product has been brought to market. This narrative contributes to the further marginalization of diverging perspectives and alternative trajectories of the technology that pointed away from the pressures of commercialization and mass production.

For the kind of inquiry that we have in mind here, perspectives from STS can be productively synthesized with social movement studies. The latter field is well acquainted with studying the dilemma faced by a social movement when it tries to have its claims and grievances institutionalized without losing itself in the process. Such a theoretical synthesis has been proposed by David Hess (2005). He coined the concept of technology- and product-oriented movements (TPMs) to describe civil society mobilizations that try to bring about social change through advocating alternative technologies and industry standards. Unlike many activist milieus, these are enmeshed in the corporate world. TPMs often have to accommodate some degree of co-optation by industrial actors for their political goals to be realized.

At the outset, it is commonplace for the goals of TPMs to be articulated in liaison with start-up firms, allies in the industry, or branches of government, whose interests happen to coincide with those of the group in question. The symbiotic relationship with these external actors brings in resources, gives credibility to the cause, and opens up networks for disseminating the message or the product. In return, the social movement provides a niche market for the company's products, as well as being a pool for beta testing, market research, and so on. As Fred Turner (2006) shows, hacker culture was founded on such an alliance, which he called "legitimation exchange." The hippie movement and the nascent information technology industry cross-fertilized. The personal computer is an offspring of the "small is beautiful" philosophy of the hippies.

Gradually, as firms systematize their interactions with social movement actors, the design and meaning of the alternative technology is disfigured. New design considerations are gradually introduced into the development process, corresponding to a realignment of the group's priorities with market demands and mass-production standards. Following on from this, the transformation gives rise to what Hess (2005) terms "object conflicts" between the TPM and its for-profit allies, and between different factions within the TPM. Conflicts revolve around the proper design and/or adoption of a transformed technology, as measured against the original grievances and values claimed by (a fraction of) the movement. External actors may weigh in behind one or another side in the internal dispute.

Arguably, the free software movement could be considered a specific case of the more general TPM phenomenon. This synthesis of perspectives from STS and social movement studies offers a terminology and a point of orientation for studying the microprocesses of resistance to recuperation within hacker projects and hacker communities. What is missing from such a theoretical synthesis, however, is a vocabulary for speaking about the systematic character of these struggles. This systematic character is due to the industry's structural dependency on hackers as an external source of innovation and value production. Hence, perspectives from labor process theory and workplace studies need to be included in the discussion.

In a branch of media studies oriented toward a political economy style of analysis, it has long been debated whether or not fan fiction writers and television audiences should be considered producers of value for the culture industry. Several attempts have been made over the years to extend the labor theory of value to include these sites of value production (Smythe

1977; Fuchs 2015a, 2015b). Alongside fans and audiences, the same kind of argument can easily be extended to more technically oriented subcultures of hobbyists, DIY-tinkerers and "citizen scientists.". The commercial breakthrough of the GNU/Linux operating system in the late 1990s and early 2000s generated much enthusiasm because it suggested the feasibility of organizing programming work with a minimum of managerial control, with the programmers self-allocating tasks among themselves. In addition to spawning numerous empirical studies of the phenomenon, it inspired conceptually driven inquiries into an emerging mode of commons-based "peer production" (Benkler 2006; Rigi 2012). Free software development showcased the failure of capital on its home turf, in organizing production in the most technically advanced sectors of the economy. Waged and hierarchical labor relations proved to be inferior to a production process based on voluntary and nonmonetized contributions from peers, not only in moral terms, but also in terms of technical efficiency (Dafermos 2012).

The dark side of this promise, which was quickly pointed out by critics of the literature on peer production, was that firms could now tap into the unpaid, collective labor of free software developers. Pressure was thus put on wages and working conditions of employed computer programmers (Terranova 2000), who are now also called the "programming proletariat" (Jordan 2016). That this warning was warranted has since been amply demonstrated by the surge in what is euphemistically known as the "sharing economy." A thin line separates the utopian promise of overturning capitalist relations of production on the one hand and the intensification of commodified and exploitative relations on the other.

The academic debate on peer production versus free labor revolves around the question: Which one of the two scenarios will prevail? Instead of attempting to settle this question as though it was an either/or proposition, we offer for consideration a theoretical synthesis drawing on the aforementioned work on TPMs. On which side a particular hacker project will end up must be decided on a case-by-case basis, depending on whether the hackers can detect and resist recuperation attempts in the local setting. To the hacker community, defeat means that its autonomy is curtailed and its collective labor is subsumed under a value-extracting, open innovation model. To everyone else, the subjugation of the individual hacker community will be experienced (if at all) as an intensified commodification of everyday life, an intensified exploitation at work, and an entrenchment of the division of labor outside of the contractual employment relation.

Using this interpretative framework, we can make sense of the regularly observed outbursts of strife and contention within hacker communities and, furthermore, explain why those outbursts tend to coincide with the crystallization of a potentially marketable product. Accusations and counteraccusations about breaches of the free/open license give a clear signal that recuperation is underway within a hacker project. The textbook example of this is when software published under a free or open license is copied by an entrepreneur or a firm and locked away behind traditional intellectual property rights or a black-boxed hardware casing (Brodkin 2016). Enclosure of the information commons by legal and/or technical means in open confrontation with the norms of the hacker community constitutes a highly visible form of direct recuperation. For the very same reason, it is often met with fierce resistance from hackers.

Inevitably, a moral register will be drawn upon by practitioners, who will accuse each other of hijacking, betrayal, and so forth. Naming and shaming are an indispensable tool in guarding the information commons from hostile encroachments. It goes without saying that an analyst drawing on the interpretative framework that we are proposing here will not be a neutral observer of events. We do not believe, however, that the predictive and analytical effort is helped by the analyst becoming an advocate siding with one or another camp. This would misconstrue the contradictions and constraints under which people subsist under capitalism. What is more, it puts the analyst at risk of being hoodwinked along with their chosen camp of practitioners, in case they are captured and subsumed under an open model of capital accumulation.

This last point is crucial, because the concept of recuperation is not exhausted by overt attempts at hijacking projects and communities. More insidious are the processes of indirect recuperation that work "behind the backs" of hackers, molding the goals and values of their community to make a better fit with the needs of the industry. Through such slow-working processes, the ground is prepared for the enclosure of the project's output in the future. This can happen without anyone even noticing, since the norms of the community are not being challenged head on. Rather, the local setting is carried away by a landslide that displaces a whole movement of hacker projects. It takes a longitudinal and comparative analysis to register such historical forces at work. The present-day project with its stated aims can be held up against the standards of its own, earlier self. Hence our insistence above on the necessity of studying the entire life cycle of a development project. Some turning points in the trajectory of a project that could warrant analytical interest include: changes in the licensing terms, terminological shifts, challenges to project leadership, and forks in the development stream.

In the case of the open-source desktop 3D printer called RepRap, mentioned above, and to be discussed in chapter 4, the growing influence exerted by the consumer market over the development process meant that waning efforts were being devoted to designing the parts for the machine in such a way that another, similar machine could print them. Coupled with the engineering goal of building a self-reproducing machine was a utopian promise, as expressed in the motto of the project: "wealth without money." The twisting of the development process around a different set of engineering trade-offs reflected a transformation in the very purpose of the project. When this happens, there are often some in the community who remain loyal to the original vision and sound the alarm. However, the losing side in an internal dispute tends to quit the community, after which the traces of there ever having been discord are obliterated and forgotten. Therefore, just as much attention must be brought to the absence of conflict as to its presence. Saying this is only to repeat a time-honored insight from the study of ideology construction and the engineering of consent.

Processes of recuperation are difficult to detect because they preserve the content but alter the form of whatever it is that has been recuperated. To illustrate this rather abstract claim, we offer for consideration the example of hackathons. This is the name given to gatherings of programmers during which they shut themselves away for several nights and days to do spurts of coding. Hackathons reflect the intense and withdrawn lifestyle of the original hacker culture (Levy 1984, chap. 4). In recent years, however, hackathons have metamorphosed into a method for brainstorming start-up ideas and for overcoming hurdles in commercial project development (Irani 2015c). Even though the practices remain exactly the same as before, the overall purpose and meaning of a corporate-initiated hackathon are very different from those of a self-organized one (D'Ignazio et al. 2016).

The lessons drawn from the case of hackathons resonate with the historical emergence of the factory as the key site of production during industrialization. Karl Marx teased out the analytical distinction between the formal and the real subsumption of labor under capital from the development of factories. Formal subsumption corresponds to an early phase in the industrial revolution when the putting-out system predominated. Merchants provided workers with raw materials and bought back the refined products at a discount. Crucially, they left it to the workers to choose their own production methods. The introduction of the factory system did not immediately deprive workers of this discretion. To begin with, the factory was just an empty building within which the workers gathered. Only gradually was the internal composition of the labor process dissolved and reorganized at the behest of capital (as documented in further detail by Thompson 1963). The human worker was transformed into a mere appendix of the factory machine, in Marx's iconic expression. In order for this machine to operate smoothly, its unruly, human cogs had to be pacified through (despotic or scientific) management.

Hacker communities are caught up in the same whirlpool of forces as factory workers once were. Hackers clash with industry (upon which they nevertheless depend as a source of income, as a provider of critical services, products, and infrastructure, and in order to gain political leverage) over the relative degree of autonomy that they enjoy in defining common goals and choosing the proper methods for developing a technology. Defeat means that the hacker community is annexed to a firm's open innovation model. Methods for asserting managerial control over the external source of value production are introduced step by step by corporations. Thus, the hacker project is turned into a more reliable site of value production for capital. Innovation is the outcome of failure to resist a recuperation attempt.

SECOND TIME HORIZON: RECUPERATION
OF THE HACKER CULTURE

The second time horizon within which we can observe the dynamics of critique and recuperation encompasses a range of different hacker projects united under a single movement that evolves in tandem with a whole branch of the industry. If the direct counterpart to the hacker project is the entrepreneur and the firm, the counterpart to a movement of projects is an industry. When we study hacking at this analytical level, we look at how a landscape of interoperability standards, embedded infrastructures, intellectual property laws, alternative licenses, and so on, all of which have since taken on the appearance of "second nature," have come into being (see Russell 2014 for an emblematic study in this vein). Recuperation processes operating within this time horizon shape the framing conditions for individual hacker projects. Rather than resulting in a single, marketable product, as is the case in the first time horizon, the interplay of critique and recuperation within the second time horizon results in organizational innovations, leading to structural reforms across several sectors and/or the birth of new industries.

It is necessary to complement case studies of hacker projects/communities with a longer historical perspective, otherwise we would lose sight of the framing conditions of those individual projects. When moving up to this level of abstraction, the analysis is conducted at one remove from the locality of the practitioners and their direct experiences of recuperation attempts. Constructivist scholars in the field of STS will object to the deterministic slant of such an argument, which they believe rules out agency. We concede that our interpretative framework does rule out agency, provided that this word is interpreted in strictly individualist terms, as the freedom of an individual to act unhindered by constraining forces. However, our interpretative framework accords with a collectivist understanding of agency, although we prefer to talk about it in terms of a political, collective subject. Indeed, the purpose of historicizing the framing conditions of a hacker project is to dismantle the appearance of (second) nature that those conditions tend to acquire over time. Looking backward, the legal and technical landscape within which hackers find themselves can be shown to be a remnant of past struggles. Looking forward, the outcome of past struggles can be shown to determine the balance of forces in the present and in struggles yet to come.

Implied in the reference to the naturalness of a second nature is that the critical reading of the situation that we put forward here will not be immediately apparent. Consequently, when working at this level of abstraction, our theoretical claims about recuperation find less support in the

testimonies of practitioners, compared to when analyzing hostile enclosure attempts within a particular hacker community. That being said, a
window of opportunity for observing recuperation processes within this
intermediate time horizon is opened up when hackers face strategic junctures with a bearing on the future of the whole movement of projects and
communities. A few examples include: the splitting of the free software and
open-source movement, the introduction of a third, updated version of
the General Public License, and the competition between different licenses
dedicated to open hardware development. However, the turning points in
hacker culture can often only be discerned with the benefit of hindsight. It
takes a long historical perspective for the slow-moving processes of recuperation to become detectable within the empirical material. Hacker histo-
riography plays a strategic role in the struggle against recuperation.

Steven Levy's classic work about hacker culture offers a point of departure for developing such a historical approach (1984). He identifies generational shifts in the history of hacking, spanning from the MIT hardware
hackers of the 1960s to the emergence of free software in the 1980s. The
common ethos, cultural references, and political goals that hackers rally
behind have undergone shifts from one generation to the next. Levy also
shows that each generational wave of hacking was closely intertwined with
developments in the computer industry and government (Zandbergen
2011), a conclusion that has since been reaffirmed by many historians. By
following the biography of Stewart Brand and his associates, Turner has
documented how the West Coast counterculture of the 1960s branched off
into a cyberculture, from which grew the nascent Silicon Valley (2006). As
mentioned above, the idea of building a small, personal computer to foster
free communication practices and new modes of community grew out of
the "small is beautiful" philosophy that underwrote the hippie communes
and the political protests that were taking place at the same time (Flichy
2007). These mutual influences went deeper than the mere exchange of
legitimacy described above. Cultural and organizational elements of the
counterculture, which already resonated with tropes from information
theory and cybernetics, were foundational for the computer industry.

Continuing the list from where Levy left off, the free software movement has been complemented in recent years by movements in physical
hackerspaces, in open hardware development, and in DIY biohacking. It

goes without saying that the reality is messier than this compilation of examples. No sharp line separates these movements, nor do they follow one another in neat succession. The free software movement has not disappeared from the map because of the rise of a movement around open hardware development. That being said, it is useful to distinguish between them analytically. Each one consists of a constellation of different hacker projects, held together by the common pool of resources and tools from which they draw. They can also be told apart by the distinct legal and regulatory spaces particular to each field. Corresponding to these are different branches of the industry. The role that the computer industry plays for the free software movement is equivalent to that of the home electronics industry for open hardware developers. Biohacking relates to various agricultural and medical branches of the industry, in addition to entertaining a special relation with US federal authorities.

From the history sketched out above, it must be clear that there never was a pristine, golden age of hacker culture that was captured at a later point by the industry and governments. The antagonists evolved in tandem from day one. Putting it differently, the starting point of hacking is always already recuperated to some degree. Saying this does not rule out the possibility that there are ideas and values in a local context that need to be defended from corporate or government involvement. In this defense, practitioners rely on "usable pasts" (Kelty 2008, 64–66), of which a key idea is that of a forlorn golden age of hacking. It is with the help of such usable pasts, projected backward in time, that hackers may construct baselines for historical comparisons and rally support behind their cause. This is most visible at times when hackers face a strategic junction in the road. One such junction came with the launch of the "open source" initiative in the 1990s, which sought to steal the leadership flag from the ideologically stringent free software movement. The schism revolved around the strategic choice of how accommodating hacker culture should become toward external, corporate, and government interests.

It is not incidental that the two sides clashed over the terms of the free/ open license. Changes in the license are key, because they lay down the conditions under which firms are sanctioned by community norms to extract profit from the collective labor of hackers. Although individuals and firms are in principle entitled to make profit from free and openly licensed goods, the reciprocal obligation to disclose information puts a de facto limit on profit maximization. The struggle over the terms of the license, and the vigilance by which those terms are then enforced in hacker projects, sends a signal about where the balance of forces is at any given point in time.

We believe it is meaningful to talk about a long-term trend in the co-evolution of hackers and industry. The interactions between them are modified by the growing awareness on both sides of their mutual dependency. The historical lessons about the computer industry's indebtedness to the counterculture have been reprocessed and incorporated into an endless array of strategies and counterstrategies. On the part of hackers, the anticipation that their projects will be assimilated by corporations feeds into their representations, values, and individual choices. The aforementioned open-source initiative is a case in point. On the part of industry and government, methods and routines are developed to render interactions with hackers more manageable, rentable, and secure.

A turning point was the general realization in the corporate world that, by making their source code public, some of the waged programming processes could be put out to hackers, something that had previously happened by serendipity. Numerous experiments with reorganizing in-house programming labor have followed, clearly inspired by the project-based and community-centered model of free software development (Auray and Kaminsky 2007; Remneland-Wikhamn et al. 2011). For instance, companies organize hackathons and open competitions to solve difficult computer problems or to find security loopholes, headhunters use free software content management systems to scout for talent, and methods of coding free software have been detached from that context and reintroduced into corporations under the label "agile methodology."

A lesson of the same import, which corporate executives already understood from the first round of battles over filesharing in the early 1990s, is that innovations and profits may also be procured from customers and users hostile to a firm's goals. In defying a firm's prerogatives to define prescribed uses for its branded products, hackers may stumble across new uses for old products, untapped consumer demand, and novel business models. The illegal status of these activities does not prevent firms from benefiting from the outcome. A case in point is distributed and anonymous data retrieval systems, which were initially developed to usurp the enforcement powers of intellectual property holders. Nowadays it is an industry standard, because it is far more efficient to retrieve data from multiple sources than from a single, centralized source. Academics in the field of innovation studies collect more such examples and offer their advice to companies on how to become even better at "harnessing the hacker" (Flowers 2008; Tapscott and Williams 2006).

The computer industry exports its production methods, along with its algorithmic services, to subservient sectors of the economy. A case in point is the platforms for involving users and audiences in firms' innovation processes on a systematic basis. With the open innovation model, peer production communities are transformed into pools of free labor. Peer production in its inverted, nightmarish form goes under the name of "the sharing economy." The truth behind this euphemism can most clearly be seen in the lower tier of the labor market. Experiments with self-managed, nonmonetized information systems that hackers undertook with utopian aspirations provide the backbone for capitalist restructuring in other sectors. Two examples will suffice. Information billboards that were set up by commuters in order to allocate empty car seats among themselves (ride sharing) bore fruit in Uber. A bottom-up initiative for coordinating travelers with empty couches in residential homes (couch-surfing) foreshadowed Airbnb. All it took was to add a cash nexus to the information service.

As suggested by the two examples above, the recuperation of hacking has consequences that extend far beyond the affected hacker community. Subsumed under an open innovation regime, hackers become a source of innovation and organizational restructuring for capital and, thus, an engine for intensifying commodified and exploitative relations everywhere in the economy. In order to catch sight of the framing conditions that precede and envelop hacker culture, the analysis needs to scale up to a third time horizon.

THIRD TIME HORIZON: THE NEW SPIRIT OF HACKING

The third time horizon relates hacker culture to capitalism as an evolving whole. From this perspective, the dynamics of critique and recuperation are seen as the motor, not only of an individual firm's marketing and innovation strategies, nor of an industry that undergoes restructuring, but also of epochal transitions within capitalism. During these epochal shifts, the predominant logic of how capital accumulates is rewritten from the ground up, affecting production, circulation, consumption, and distribution. It is within this third time horizon that the first (the life cycle of an individual hacker project) and second (the coevolving landscape of hackers and industry) acquire their determinate meaning. When we move up to this level of abstraction, explanandum and explanans are reversed. Instead of explaining hacking with theories about capitalism, it is capitalism that is examined through the lens of hacking.

We are indebted to Luc Boltanski and Eve Chiapello's *The New Spirit of Capitalism* in making this suggestion (2005). The book title alludes to Max Weber's famous argument that capitalism was made possible by the ethical foundations created by Protestantism. Extending this idea to modern times, Boltanski and Chiapello argue that the transitions that periodically take place under capitalism are triggered by the critiques that are leveled against it. In order for the economic system to operate smoothly, it needs to appear legitimate and even inspiring to both workers and consumers. Lacking an ethical or affective source of its own, capital must draw legitimacy from external sources. The incorporation of critique through a period of restructuring allows capitalism to overcome its own impasses.

This transhistorical scheme neatly fits the current spirit of capitalism, which Boltanski and Chiapello name "connexionist." They trace the latest iteration of capitalism back to the anticapitalist critique that emerged from the political upheavals of the 1960s. The values associated with May 1968, individual freedom, self-expression, nonconformity, authenticity, and so on, were quickly depoliticized and transformed into an ethical foundation for capitalism in its present-day, consumerist, and financialized form (see also Liu 2004; Harvey 2005).

The 1960s and 1970s have been designated a turning point in capitalism by many otherwise unrelated schools of thought. Often, the argument is mapped onto the transition from Fordism to post-Fordism. Under Fordism, mass production was the defining trait of every economic activity, including such phenomena as mass party and mass union activism. Under post-Fordism, the flexible, just-in-time production model saturates every corner of society, not even sparing cultural expressions or political activism. Approaching it from a structuralist perspective, the regulation school spoke of these epochal shifts as "regimes of accumulation" (Aglietta 1976; Jessop 1995).

The calling card of the structuralist approach is a deep suspicion directed toward the cognitive capacity of practitioners to make sense of the situation in which they find themselves. For most of his career, Boltanski has polemized against such a demeanor. Boltanski and Chiapello are aware of the fact that, by proposing a transhistorical scheme for interpreting practitioners' behavior and claims, they end up vulnerable to the same kind of accusations that they previously directed against structuralist sociology in general, and Pierre Bourdieu in particular: "Stressing historical structures, laws and forces tends to minimize the role of intentional action. Things are what they are. Yet the critical approach becomes meaningless if one does not believe that it can inflect action, and that this action can itself help to change the course of things in the direction of further 'liberation'" (Boltanski and Chiapello, 2005, x).

Undoubtedly, this warning is also applicable to the interpretative framework that we are developing here. Our musings about critique and recuperation within the third time horizon find scant support in the testimonies of hackers.

At this point in the argument, we take our cue from the autonomist Marxist tradition to explain how historical reflection on a grand scale can be developed without the analysis succumbing to deterministic explanations. Autonomist Marxists conceptualize the transition from Fordism to post-Fordism as a change in the "technical and political class composition." The stress is placed on class struggle as a driver of capital's restructuring processes. The autonomist Marxists apply this line of reasoning directly to the instruments of production. This we consider to be an advance over Boltanski and Chiapello's scheme. The latter have surprisingly little to say about the role of technology in their interpretative framework. Perhaps they dodged the question out of concern that they may become associated with the genre of writing that hails the coming of "postindustrialism," the "information age," or "network society." A common feature of this genre is that information technology is designated as the motor of historical development. With a focus that locates the root cause of technological change in the struggle between labor and capital, we can reconstruct long time horizons without having to resort to the notion that history is propelled by an "innate trajectory of technology." Furthermore, empirical support for the autonomist Marxist claim is easy to come by in hacker culture. As we discussed previously, a long list can be compiled of disruptive innovations in information systems that emerged from contestations between hackers and their corporate or state adversaries.

We are not the first to suggest that the new spirit of capitalism can be found in a condensed form within hacker culture. Drawing directly on Max Weber's writings, Pekka Himanen (2001) argues that hackers embody a new work ethic that is about to supersede the protestant work ethic of the past. He praises this trend in what amounts to a new take on an old trope in management literature, according to which the next industrial revolution will abolish alienated and deprived working conditions. Anne Barron (2013) also points to hackers as representatives of a new work ethic, but she argues in a more critical vein, starting from Boltanski and Chiapello's reinterpretation of Weber. We concur with her that the values embodied in free software development qualify as a particularly pure form of the new spirit of capitalism. The reverse side of the critique against proprietary software and other forms of closed innovation systems is an emotional investment in "open" forms of capital accumulation. Thus, the oppositional stance of hackers has been turned into an ethical foundation for contemporary capitalism (Barron 2013).

The subjective counterpart to the open innovation model is the outsider. Identification with the position of the outsider goes to the heart of hacker culture. The same idea of freedom as having "no strings attached" underpins the promise that emancipation will flow from the repurposing of tools and the circumvention of social constraints by technological means—that is to say, from hacking. Alas, such acts of technology-mediated transgression have always already been anticipated by the open innovation model. As sources of value production external to the firm become increasingly important for capital, so the more pervasive the methods will become to reassert management control over this distributed chain of production. Furthermore, open innovation is only a subcategory of a more general trend in which value production is relocated from contractual employment relations to peer production communities, crowds of users and audiences, and clouds of click workers. This is the spirit of capitalism that is embodied in the work ethic of the hacker.

THE FUNCTIONAL AUTONOMY OF HACKER PROJECTS
AND HACKER COMMUNITIES

The outcome in the struggle against recuperation decides to what extent hackers will shape technology and, conversely, to what extent hackers themselves will be shaped by and contribute to predominant structures, including predominant technologies. Struggles over recuperation center on the functional autonomy of hacker culture vis-à-vis its exteriority. By the concept of "autonomy," we understand the ability of a collective to give itself its own laws and future-oriented goals. Autonomy only exists in relative measure. The notion of absolute, unconditional autonomy is a misnomer, just as much as the fantasy of the free-floating position of the outsider, as discussed in the paragraphs above. Hence, we aspire to give equal weight, on the one hand, to the dependence of hackers on preexisting structures (such as, for instance, industrial standards, economic relations, cultural values, etc.) and, on the other, the discretion that hackers nevertheless possess to pursue pathways in technology development that diverge from predominant trends.

We build on a tradition within political philosophy in which autonomy is understood as a phenomenon that first emerged in struggles for political emancipation. Only much later was autonomy defined in analytical-philosophical terms. The point zero of autonomy is not, in other words, Kant's philosophy, although he is an important stopover in the modern history of the concept. Rather, the idea of autonomy dawned on humankind during the process of resisting imperial and dynastic domination. Two milestones in this history are ancient Greek poleis and city-republics during the periods of the Renaissance and Reformation, who mobilized against oligarchic coups and armed interventions. From this, there follows a point of utmost importance. Autonomy is not a mark of individual cognitive and moral beings, but rather a property of collective, political subjects that is consolidated through struggle (Rosich 2019).

Likewise, we take issue with an influential tradition in political philosophy, represented by Hannah Arendt among others, according to which "autonomy" is a sphere of freedom that is antithetical to needs, work, and technology. If autonomy is conceptualized as ontologically opposed to labor, then one makes oneself oblivious to contestations over autonomy that are fought out on a terrain of material needs, economic dependency, and technological infrastructures. In the idealist school of political philosophy, "autonomy" is understood in such a way that it rules out the possibility of an "extension of the franchise" to the working class. This philosophical outlook resonates with millennia-old class prejudices.

As a corrective to both the individualist and idealist readings of the word "autonomy," we draw inspiration from how struggles over autonomy have been conceptualized and studied in labor process theory (Hanlon 2016). In case studies of worker resistance against management prerogatives and Taylorism, it is a given that the autonomy of workers is limited by the contractual employment relation in which they find themselves. This contrasts with the abstract notion of "absolute freedom." Furthermore, in the context of workplace struggles, autonomy must be understood as embedded in a material substratum: the distribution of skills among the workforce, the layout of the factory and design of the machinery, the economic bargaining power of the class antagonists, and so on.

From David Montgomery's seminal study of labor activism during the second half of the nineteenth century, at a time when factory operations were largely in the hands of self-governing craftworkers, we derive the notion of "functional autonomy" (Montgomery 1987). Due to their intimate familiarity with the production process, craftworkers exercised a functional autonomy over the factory, even though, nominally speaking, that same right to dispose of their time and effort was held by the factory owner. It is precisely this discrepancy between the nominal rights assigned to the capitalist, and the practical control exercised by the craftworkers, that transformed skill levels and machinery design at the point of production into a major battleground in the ensuing cycles of struggle.

The continued relevance of investigating the functional autonomy of workers, however much it has been encroached upon compared to that of nineteenth-century craftworkers, is attested to by the battery of countermeasures deployed by the representatives of capital, such as, for instance, scientific management, the automation of work tasks, and union-busting tactics (Delfanti 2021). From labor process theory, we derive an analysis of technological design and skills that are understood to be through and through determined by the balance of forces between labor and capital.

Technology is pivotal in this interpretative frame, but not exclusively so. We have also taken a cue from labor process theory in directing our analytical attention toward the intersection between artifacts and skill, on the one hand, and group solidarity and norms, on the other. The two are equally decisive for the outcome of industrial conflicts (Montgomery 1976).

The tug-of-war between workers and managers over who is in effective control over the shop floor provides a blueprint for studying the functional autonomy of hacker culture within a relation of dependency upon the industry. Although, by definition, hackers have no contractual ties to an employer, they are nevertheless dependent on the industry in numerous respects: in order to be paid commission, access technical products, gain political leverage, and so on. For the hacker, just as for the employee, the distribution of productive skills and the design of machinery are decisive factors. The autonomy of hackers is curtailed, for instance, when software tools and software libraries are fenced in behind intellectual property rights. Conversely, their autonomy expands as tools and libraries are brought into the information commons. Whatever else is being produced by a thriving, autonomous hacker project, it concurrently furnishes the material conditions for its own continued existence as an autonomous group entity.

We label a positive outcome in struggles over recuperation "communization." We take this word from the discourse of the self-described ultraleft tendencies, where it refers to steps toward the direct realization of Marxian communism within the context of everyday social relations (e.g., Dauvé 2011; Friends of the Classless Society 2016). This is in line with our conception of the term, but we use it in a narrower sense here, referring to the rerouting of work time and resources, nominally owned by the employer, to sustain commons-based peer production communities. The resources may also be procured through sponsorship from companies. Even so, it indicates that the hackers are in a strong position to define the terms of engagement with industry actors. In chapter 6, we discuss Internet Relay Chat as an example of communization. For decades on end, system administrators working for internet service providers and university departments have allocated their employers' servers to run public Internet Relay Chat networks. Their ability to do so, with or without the tacit agreement of line managers, hinges on them having technical know-how and access rights. Hence, processes of communization often interlock with a high degree of functional autonomy.

A bone of contention in labor process theory, which has a bearing on our discussion about hackers, concerns the double-edged implications of workplace autonomy. Coercion is not the sole tool in the manager's toolbox. Another way for managers to ensure consent to work is to give workers some leeway (Burawoy 1979). Indeed, it is commonly observed in this literature that the smooth operation of a factory presupposes that the workers have some discretion over task selection and that they feel somewhat responsible for the quality of their output. When managers tighten their control over the production process, often in response to an outbreak of industrial conflict, productivity tends to be negatively affected.

The same qualification applies to the subsumption of a hacker project under an open innovation model. The reason why capital relocates production from employed, in-house labor to commons-based peer production communities is not merely to tap into free (as in gratis) labor. Equally important is the higher productivity and greater inventiveness that is fostered outside of the wage labor relation. As much is suggested by the technical superiority claimed for GNU/Linux over proprietary operating systems. The coerciveness of waged employment is suboptimal for organizing productive activities, at least in the upper segment of the value chain, where value is extracted from the workers' subjectivity. The open innovation model seeks to remedy this situation. Managerial control over the innovation process is weakened in order to set productivity free. That being said, some degree of managerial control must nevertheless be upheld, in order for capital to valorize the collective endeavors of hackers. Furthermore, the imperative of profit maximization spurs firms to reassert control over commons-based peer production communities, occasionally killing the proverbial goose that lays the golden eggs.

As has already been established in labor process theory, there is a tension between the need for capital to, on the one hand, assert managerial authority over the labor process and, on the other, give workers some margin of freedom in order for them to determine the most productive application of the factory machinery. This tension is typically resolved by refining the division of labor among the workers. Tasks are unevenly distributed among different classes of workers, a separation reinforced by status hierarchies at multiple levels (education, gender, ethnicity, and nationality, to mention the most important ones). Building on this historical lesson, we suggest later in this chapter that there is an analogous development taking place outside of the contractual employment relation. A division of labor is emerging between three different classes of nonemployees: communities of peer producers, crowds of users, and clouds of click workers. We base this classification scheme on the relative degree of functional autonomy that is exercised (or not) by these collective formations.

Obviously, hackers sit at the top of this value chain. By saying this we have not, however, exhausted all that can be said about them. Rather, this observation serves as our starting point for investigating the shifting degrees of autonomy of different hacker communities, all of them privileged in comparison to the click workers. In our understanding, it qualifies as a high degree of autonomy when hackers can dictate the terms under which they (collectively and individually) enter into a symbiotic relationship with industry and government actors. On this hinges their ability to cultivate critical opinions and imaginaries about technology that deviate from mainstream engineering practices. In the absence of such autonomy, hackers will develop technology that reproduces predominant structures and proclivities. Concurrently, however, in order to gain traction, hackers must produce something of value to society, which is what gives them leverage against their more powerful allies. In the absence of such allies, the hacker project will become isolated and lose its societal relevance. The symbiotic relationship between hacker culture and the computer industry is not optional—the real question is on whose terms this symbiosis is based.

In order to speak about functional autonomy with more precision, we now proceed to identify three pillars upon which it rests. The first pillar is the technical expertise of hackers. The second is their shared values and norms, cultivated and sustained in partial isolation from mainstream engineering culture and society at large. The third is historical memory, or, put differently, a common narration of events in the past with a bearing on the collective's future development. All of these pillars must be in place for hackers to successfully reproduce the social and material conditions for their own continued existence as an autonomous and self-directing collective. On this, in turn, hinges the capacity of hackers to pursue alternative pathways of technological development, as well as their cultivation of independent opinions about policymaking related to information systems.

TECHNICAL EXPERTISE

Hackers' ability to understand and produce technology is important in a number of respects. Understanding how something works is a prerequisite for judging its wider significance to one's community and society at large. Such judgments inform and enable critical engineering practices, by which we are referring to hands-on practices and design choices that seek to promote social relations different from the hegemonic ones. Hacker politics chiefly consists of the creation of artifacts (software, hardware, protocols, or biological processes). It is on this score that, similarly to TPMs more generally, they distinguish themselves from traditional social movement activism, such as street protests and the petitioning of politicians. Furthermore, the spread of hacker culture lowers the threshold for accessing technical expertise. In the best-case scenario, disenfranchised groups may thus acquire the necessary know-how for contesting issues of immediate concern to them, above and beyond the information systems that hackers care about.

The same skills that circulate in the computer underground are also on offer on a regular engineering curriculum. Even so, these different training grounds give rise to very different results. The prescribed career path of the engineering profession is, firstly, to have been a hobbyist in one's teens, then to get a university degree in informatics or physics, and, finally, to get a job in a tech company or start one's own firm. Many computer engineers walk this career path without ever coming into contact with hacker culture. The educational system gives them the competences to assess and intervene in technology, which amounts to the first pillar of autonomy in our analytical scheme. However, having been drilled in a hegemonic conception of what technology should be, professional engineers tend, even in the absence of financial incentives and managerial dictates, to channel their efforts into the reproduction of predominant social relations (Noble 1977). The criteria for deciding the success or failure of a development project imbued with mainstream engineering culture is laid down by "instrumental rationality." That is to say, the whole endeavor will be directed toward the optimization of cost efficiency relative to technical performance.

Hackers are not oblivious to performance in the narrow, technical sense, but aesthetic and ethical considerations also carry great weight when they assess whether or not to adopt a new artifact. A premium is often placed on simplicity and transparency, which occasionally overrule the imperatives of cost efficiency, speed, and user convenience. This can be observed in the common practice among hackers of sticking with outdated and, in terms of technical performance, "inferior" products and standards, in comparison to the up-to-date versions. Noteworthy is the widespread rejection (Wyatt 2010, 9) and critique of smartphones, commonly referred to in hacker lingo as "tracking devices." Just by being slightly out of sync with the latest marketing scheme, outdated technology may give its users some extra leeway. Furthermore, the slower pace of diffusion of older technology is conductive to its integration into moral economies, social conventions, and norm systems that have not been fully subsumed under corporate and elite control. As the gap widens between the old standard and the latest iterations, however, inoperability takes its toll, and the alternatives promoted by hackers become isolated from the wider society. The goals of ideological purity and political relevance are in tension. Hackers must refine their taste in judging the right balance between the two goals and realizing when to switch platforms, along with reproducing more conventional skills in soldering, programming, and so on.

SHARED VALUES AND NORMS

Cultural tropes and values are reproduced where hackers gather. The most frequented of such meeting places are the ubiquitous asynchronous and synchronous online social spaces, such as mailing lists and chat rooms (the latter analyzed in chapter 4). Complementing online interactions, physical meeting places serve a critical function in disseminating and entrenching a specific hacker culture. Hacker conventions—dubbed by Coleman "a ritual condensation and celebration of a lifeworld" (Coleman 2010)—allow hackers to convene at regular intervals from many different places, while shared machine shops provide fixed spatial coordinates for the hackers in a city or region to meet and forge bonds over time.

Shared machine shops, notably the hackerspace, as described in more detail in chapter 3, are a materially and symbolically constituted milieu of hacker culture. In this setting, subcultural symbols and rituals are transmitted via everything from culinary preferences to references from popular culture. Some examples include: the fridge in a hackerspace is typically stocked with Club-Mate, the default drink of hackers (Thomas 2014); the soldering iron stall warns that "if it smells like chicken, you're holding it wrong," which is a reference to Mitch Altman's soldering workshops; names of people and artifacts are taken from Discordianism, the Cthulhu mythos, or *The Hitchhiker's Guide to the Galaxy*. Thus, specific tastes, habits, and even metaphysics are reproduced in hackerspaces that lend support to and legitimize the core technical practices. Hackerspaces, although varying greatly in cultural and ethical preferences, constitute the most stable physical manifestation of hacker culture.

Technical expertise (the first pillar) and shared cultural tropes and values (the second pillar) will not on their own suffice to sustain the functional autonomy of hacker culture. As much is suggested by the Asian and Chinese hacker scene. From the research literature, we learn that "hacking with Chinese characteristics" draws on the same skill set and roughly the same cultural tropes as those circulating globally, but the antiauthoritarian and confrontational outlook of the original hacker identity is largely absent (Lindtner and Li 2012). The same observation can be made about the geographical displacement of shared machine shops, from Europe to North America and then to every corner of the world. The anarchist politics of the original hacklabs was lost in translation when the idea caught on in the United States and they were rebranded as "hackerspaces" and "maker-spaces." This prompts us to stress the third pillar upholding the functional autonomy of hacker culture: the transmission of shared memories and lessons learned from the defeats and victories of older generations, which can provide points of orientation for future-oriented, collective action.

HISTORICAL MEMORY

Familiarity with past waves of technology is passed on to new hackers in part through the floating debris of obsolete gear that piles up in most hackerspaces. Recycling is the backbone of their political economy, the junkyard furnishing them with both spare parts and inspiration for new projects. Antiquated machines, unfinished projects, and random electronic parts are stored on the shelves of hackerspaces and serve as a reference library of engineering solutions. When wondering about the proper way to wire a chip, one can find such chips already wired into some other device and use it as an example. Aside from the pragmatic aspects of recycling, old artifacts also elicit sentiments of nostalgia and even veneration. Some hackerspaces end up playing the part of technology museums. A case in point is Hack42 in Arnhem, the Netherlands. Housed in an old military barracks modeled after German countryside cottages, the three-story hackerspace includes several thematic collections of obsolete hardware: cameras both analog and digital, overhead projectors and beamers, typewriters, calculators, and computers, many of which are kept in working order.

Another anecdote suggestive of how old artifacts are called upon to articulate a critique of current trends in technological development is Mitch Altman's signature invention, TV-B-Gone. This is a universal remote control with a single button that turns off any television. Hackers wielding the TV-B-Gone convey the antitelevision sentiments of an earlier generation of computer users, for whom the television symbolized passive media consumption bordering on corporate mind control. The rejection of mass media was a thematic core for the hardware hackers who first envisioned the "small is beautiful" personal computer in the 1970s (Levy 1984). At the present moment, however, the TV-B-Gone device doubles as a critique of the ongoing convergence between the internet and broadcasting media, culminating in corporate-controlled video streaming platforms.

Appreciation for bygone computer architectures and artifacts sensitizes hackers to the alternative pathways that technology could have taken under different circumstances. It provides them with a baseline for comparison with actual developments in information systems. Nostalgia serves as a much-needed antidote to the presentism that is rampant in the high-tech sector. Tied to this backward-looking sensibility is a diagnosis of the actors and structures that impinged upon the development process in the past and, hence, continue to do so today. The interpretation and transmission of past events with a bearing on the collective identity of hackers is key to shaping their political outlook and guiding future-directed, concerted action.

COMMUNITIES OF PEER PRODUCERS, CROWDS OF USERS, CLOUDS OF CLICK WORKERS

Thus far into our description of how hackers become caught up in struggles over recuperation, we have postponed the crucial question of if, and if so, how, solidarity across petrified and compartmentalized identity boundaries can be constructed on the basis of hacker culture. We have no illusions about this being an easy task that is just waiting to happen. Both subjective and objective factors pull hacker culture in the opposite direction. Still, unless such bonds of solidarity can be forged with other social movements and disenfranchised constituencies, the hacker culture is destined to become a subservient incubator of innovation for the benefit of capital.

In terms of the subjective side of hacker culture, the political agnosticism of hackers is renowned (Coleman 2004). This does not rule out activism, but it does put an inward spin on their version of politics. As we discussed above at length, the orientation of hacker culture is toward expanding the material conditions for perpetuating its own existence as an autonomous collective. In the literature, this trait of hacking politics has been described as "recursiveness" (Kelty, 2008). The advantage of recursive politics is that, when hackers take a stand on policies that have a bearing on their continued existence as a collective, for instance, regarding extensions of intellectual property rights or deviations from the principle of net neutrality, they act in concert. The downside of the same thing is that hacker culture tends to be unreceptive toward larger political issues that do not arise from within this recursive loop. A case in point is the coarse welcome given to female hackers in some discussion forums and hackerspaces. The first kind of topics are typically construed as apolitical, a mere optimization of given engineering parameters, even when applied to policymaking and legislation (Gillespie 2006). The second kind of activism comes across to many hackers as ideological foul play by intruders. In sum, the political agnosticism of hacker culture does not promote the forging of solidarity bonds that extend beyond its own, insular concerns.

To the subjective side of hacker culture, we must add the objective class position of the computer engineer, to which the hacker belongs at one remove. Without question, the engineering profession is one of the most privileged segments of that part of the population that has to earn its means of subsistence on the labor market. In addition to the high salaries in the high-tech sector, engineers are overwhelmingly white, college-educated males living in metropolitan areas in the Global North. From a world system perspective, programming labor sits at the top of a global value chain that descends to miners who excavate lithium in Bolivia, maquila workers producing home electronics in Mexico and China, migrant workers in metropolitan tech centers providing supportive functions, and slum dwellers around the planet recycling electronics waste (Dyer-Witheford 2015). At the point of production, finally, computer engineering can serve various functions, but the decisive one is to reinforce management proclivities. The history of computer programming goes back to numerical control machinery that was deployed by managers to extend their control over factory machinery, and, by extension, over the machine operators (Noble 1977).

At best, it may be granted that computer programmers belong to a modern version of the "labor aristocracy." If so, two partially countervailing lessons can be teased out from the history of organized labor. The first lesson is that with privilege comes susceptibility to hegemonic ideas and values. One may easily draw parallels between, one the one hand, the exclusionary practices and macho jargon that saturate much of hacker culture (Bardzell, Nguyen, and Toupin 2016; Dunbar-Hester 2019) and, on the other, the strategies by which craftworkers once protected their working conditions by excluding female and casual laborers from workplaces and/ or union membership (Cockburn 1985). There is, however, another story to be told about the labor aristocracy. Spared from absolute destitution and precarity, they were in a better position than many other elements of the working class to organize themselves and their fellow workers into labor parties, trade unions and consumption cooperatives (Moorhouse 1978).

Extrapolating this historical lesson to present-day hackers, we note that the autonomy of hacker projects is sustained in large part thanks to the resources that are pulled into the hacker movement by overpaid programmers. Furthermore, the alternative design choices made in this milieu, such as, for instance, the decision to make source code public, is beneficial to the majority of computer users, although they will never engage directly with computer programming. The point is that there are contradictory potentialities contained within the subject position of the hacker. Which of these potentialities will be actualized is decided over the course of an ongoing struggle, hence the outcome cannot be told in advance. How we choose to study hacking contributes in some small measure to this outcome.

Academics like ourselves who criticize hackers for being technophilic and for engaging in exclusionary practices have a point insofar as those critiques are made with the intent of encouraging hackers to incorporate previously disenfranchised groups. The interpretative framework that we are developing in this book, in which hacking is situated within a larger whole of evolving capitalist relations, is not meant to be explanatory only. By connecting the lines between seemingly unrelated dots, we want to suggest that the "loop of recursiveness" that delimits the agnostic politics of hackers ought to be cast much wider. Implied in this analysis is the conclusion that hackers strengthen their own position by extending the bonds of solidarity to other groups who are caught up in the same forces of recuperation.

In order to continue along this train of thought, we must zoom out from the special case of hacker culture and revisit the discussion from the opposite direction, that of the spirit of capitalism. The open model of capital accumulation is mirrored by a working class that, even after the factory walls have been demolished and the collective identity of the Fordist mass worker has been dissolved, is still obliged to sell its labor in order to earn a subsistence. It is therefore urgent to investigate the subjective experiences of class within a dispersed, capitalist chain of production with the following question in mind: What collective representations can emerge from a setting where the means of living must be earned on the labor market, but where the everyday experiences of class antagonism are no longer framed by the bipolar conflict of interests between employer and employee, as codified in the employment contract? Employer and employee confront each other over working hours, pace of work, remuneration levels, and so on. Such pedagogical support is missing in the new forms whereby capital extracts value from labor. With the exception of a handful of rare cases (Postigo 2004), fans, gamers, users, and so on, even though their hobby has become integrated into capital's circuits, continue to refer to themselves as something other than exploited workers (Lee and Lin 2011). The fragmentation of class antagonism is fueled by the never-ending stream of neologisms invented by the platform owners to name their workers, such as "taskers," "runners," and "Turkers" (Irani 2015a). The argument comes full circle when we insert the "maker" and "hacker" into this context. Hacker projects are not only a showcase for how commons-based peer production communities can be put to work by firms. Once they have been put to work in this way, hackers make substantial contributions to the organizational and material infrastructure of the emerging social factory.

When a hacker project has been subsumed under an open innovation model, it furnishes capital with ideas and innovations that can then be deployed in other sectors of the economy. At first, it might seem as though those who stand to be most negatively affected by the free offerings of programming labor would be waged labor in the same sector. However, the salaries and working conditions of waged computer programmers have not been markedly affected by the surge in free software development. The case could even be made that their collective bargaining position has been strengthened thanks to there now being an alternative forum and labor market to which they can turn. Ultimately, this comes down to them acting from a position of strength. The objective class position of programmers in the global and social division of labor is such that they stand to benefit from the transfer of resources to the high-tech sector from all the other sectors of the economy.

It is in the lower tiers of the labor market that the coercive side of recuperated engineering utopias come to the fore. Euphemisms such as "the sharing economy" and "the gig economy" cast a veil over the exploitative and precarious working conditions of a new generation of workers (Scholz 2016). This is the mirror side of capital's growing dependence on sources of value production external to the firm and, consequently, external to the contractual employment relation. With this comes the need to assert managerial control over the decentralized labor process. Corresponding to this development, a new division of labor is cropping up between different classes of developers, beta testers, users, audiences, and, further down the chain, taskers, Turkers, runners, and so on. We propose the following typology: "communities of peer producers," "crowds of users," and "clouds of click workers."

At the pinnacle of the open innovation model is the commons-based peer production community, of which free software developers are the paradigmatic example. This community is a self-initiated and voluntarily entered association with a large capacity for collective action. Autonomy is a necessary condition for incubating the kind of innovative and problem-solving activities from which capital derives the highest value. Next in line come the crowds of users and audiences of various sorts. They swarm together on what seems to be a voluntary (noncontractual, nonremunerated) basis. Upon closer inspection, however, it turns out that they have often been algorithmically herded into environments where their activities can be mined for data or for other kinds of long-tail derivatives (beta testing, computing power, etc.). The unpredictability of this extraction model is that, under exceptional circumstances, the crowd may explode with political energy and, at least for a brief moment, turn into an angry mob. At the bottom end are clouds of click workers, who perform predefined, routine tasks on corporate-owned digital platforms under strict surveillance. Just as with the layout of the factory, the digital platform has been conceptualized from the outset to minimize communication and self-initiated coordination among the individual members of the cloud.

The list "community, crowd, and cloud" is made up of names that we give to different, nonemployed work constellations. These words have long-established uses and connotations predating the rise of information systems. What motivates our chosen terminology, however, is the common usage of these words in the context of online communication technologies. In keeping with the classificatory work previously undertaken in social movement studies, we distinguish community, crowd, and cloud according to their differentiated capacity for collective action (Dolata and Schrape 2016), while adding an additional layer to this analytical scheme by considering how capital extracts value from them. Hence, the defining criterion of communities, crowds, and clouds consists of how much discretion (if any) these constellations exercise over the purpose to which their labor is put. The relative degree of functional autonomy enjoyed by these constellations stands in an inverse relationship to their subsumption under capital. The cloud of piece rate click workers offers a near perfect image of what the real subsumption of labor under capital looks like outside of the contractual employment relation. The commons-based peer production community, in contrast, showcases a highly autonomous and self-directed labor force, to the point where one may easily forget the influence of capital and capitalism in such a setting. It is for precisely this reason that a theoretical reconstruction of the structural interdependencies of these different work constellations is called for. In short, autonomy is a precondition for communities of peer producers to furnish capital with disruptive ideas and innovations, so that capital may better subjugate the crowd and the cloud (together with the regularly employed workforce, of course) under managerial control, consumer surveillance, and intensified exploitation.

This bleak scenario is not the only one possible. The legacy of hacking contains contradictory potentialities, some of which point toward a broader political-economic analysis and the forging of solidarity bonds with the working class at large. We are reminded of this possibility by the concluding words of one of the founding documents of hacker culture, the GNU manifesto: "We have already greatly reduced the amount of work that the whole society must do for its actual productivity, but only a little of this has translated itself into leisure for workers because much nonproductive activity is required to accompany productive activity. The main causes of this are bureaucracy and isometric struggles against competition. Free software will greatly reduce these drains in the area of software production. We must do this, in order for technical gains in productivity to translate into less work for us" (Stallman 1993).

3

COMMUNITY WIRELESS NETWORKS
A DARKNET OF LIGHT

Almost ten years prior to the emergence of the global open hardware movement, a local open hardware project called "Ronja" flourished in the Czech Republic. Ronja was a mechanical device for connecting computers point to point with visible, red light. The technical term for "blinking" data is "free space optics" (FSO). The home-built FSO technology provided the backbone for many community Wi-Fi networks in central and eastern Europe during the early 2000s. The wider political aspirations invested in the diffusion of this communications technology are nicely captured by an anecdote about a minor deviation in the design of Ronja.

Many of Ronja's users, often teenagers and students, lived in communally governed tower blocks. Their older neighbors did not always appreciate the light show on the rooftops. Being in control over the local housing committees, the neighbors had the authority to tell the young residents to take down their devices from the building. This happened frequently enough to encourage the participants in the wireless network community to investigate a technical solution to the problem. They developed a modified version of Ronja that henceforth was called "Inferno." It differed from the original design only in that the data was transmitted in the infrared rather than the visible region of the electromagnetic spectrum. The technical performance in terms of speed and interference from rain or fog was equivalent. The advantage of operating in the infrared spectrum was that it reduced the interference from neighbors. The point of contestation had, literary, been removed from sight. Data continued to flow behind the backs of the neighbors (Sykora, November 27, 2008).

The above anecdote is suggestive of a key theme in this chapter. Design choices for technology have a bearing on what layers of reality are rendered transparent or opaque, and to whom. Transparency depends as much on the know-how of the user as on the design of the technology itself. Given that technical knowledge is unevenly distributed across society, epistemology turns into a source of power over others. It would not be far-fetched to describe the history of technology as a record of potential social contestations that were rendered unrepresentable through architectural design. In the default scenario, however, the strategic capacity to resolve political conflicts through decisions about architecture is assumed to be reserved for a ruling technocracy. The promise of Inferno, of Ronja, and, indeed, of hacking in general is that the capacity to route around political constraints and incumbent powers by way of invention can be reclaimed and put into the hands of everybody. This is the idea underpinning the claim on the official Ronja website that the FSO link was a "user-controlled technology."

No doubt, the housing committees of the communal tower blocks represented a minuscule power in the larger scheme of things. The anecdote gains somewhat in gravity, however, when recalling that the bulk of the data carried by the light beams consisted of pirated music, films, and video games, shared in violation of Czech and international copyright laws. Indeed, the stated goal of the Ronja project was precisely to build a network infrastructure that would be able to evade law enforcement agencies and other kinds of government regulation, state censorship, and corporate surveillance. One advantage of streaming communication through visible, and sometimes invisible, light, a point that was repeatedly made on the Ronja discussion forum, was the absence of regulation over this end of the electromagnetic spectrum. Ronja was, so to speak, a "darknet of light."

BRIEF DESCRIPTION OF THE DESIGN OF RONJA

A Ronja link consists of two technically identical devices mounted in line of sight of each other. The main part of the device, the so-called "head," is built from two metal chimney pipes. One of the pipes holds an LED lamp of the sort normally used in traffic lights. A lens placed at the end of the pipe focuses the beam. The other pipe contains a light-sensitive photodiode that receives the incoming light signals from the device mounted at the opposite end. The incoming light pulses are transformed into electrical charges and then into a signal that can be read by the network card in the computer. With this contraption, data can be sent at a speed of 10 Mbps over a maximum distance of 1.4 km. A major advantage of FSO compared to Wi-Fi technology is that the former can send and receive data simultaneously. Another advantage over Wi-Fi is that any number of FSO links can be operative in the same area without resulting in crowding effects. A drawback of the technology is that it is sensitive to poor weather conditions, especially fog, and other kinds of mechanical interference that block the line of sight.

The performance of Ronja, in terms of both speed and reliability, was far superior to that of any commercial equipment available in the Czech Republic at the time. It was also much cheaper than any comparative alternative. The price of a commercial Wi-Fi access point amounted to more than seven hundred euros at the time when the Ronja project was launched. The parts for building a Ronja link cost between thirty-five and one hundred euros (Sykora, November 27, 2008). Interest in the Ronja project soared for a number of years. There is no way to know how many of them were built. The Ronja website lists 153 photographed installations, but most users never bothered to document their devices or report back. A rough estimate of the number of devices that were in use can be based on a key component of the machine. The suitable type of LED lamp for emitting the light was shipped from the United States in packs of 120. A member of the Ronja community ordered packages and sold individual pieces at cost to other participants. He recalls having distributed more than eight hundred LEDs, and he is aware of others in the community who also acted as distributors on a lesser scale. In addition, large wireless network communities ordered their own packages directly from the United States. It gives some indication of the spread of the project in central and eastern Europe (Tesar, October 5, 2008). An estimate of the number of Ronja links that have been built worldwide can be derived from the centralized distribution of LED lamps, together with reports from retailers about other critical parts, such as one vendor of printed circuit boards who claims to have received roughly one thousand orders in total (Horky, January 17, 2009).

Based on these accounts, we find it plausible that a couple of thousand Ronja links were in operation at one time or another. Furthermore, since the FSO device was often used as a backbone for wireless networks, the number of computers that have been connected through a Ronja link is much larger still. The semicentralized distribution of LED lamps also gives some indication of the diffusion of the project. By far the most requests came from within the Czech Republic and, secondly, from neighboring Slovakia, but there were also users in Serbia and Romania who placed orders for the item. Difficulties in sourcing the right kind of LED in Kerala, India, resulted in a modification of the original design (Krishnan, October 17, 2008).

The limited diffusion of Ronja, even to neighboring countries in central and eastern Europe, is largely explained by restricted information being available about the project. During the first three years, it spread by word of mouth. The Ronja website was set up in 2003, but a lot of the documentation and most of the discussions were initially in the Czech language. Directed effort was put into translating the text into German and English. Language does not explain everything, however. Even in neighboring Slovakia, where there is no language barrier, diffusion was much slower. The reason for this, according to one developer living in Bratislava, was the absence of a community of users who could pool resources and assist less experienced users in building Ronja devices (Hecko, December 17, 2008).

The construction of a Ronja device took perseverance, effort, and a lot of spare time. Just finding the parts was a challenge. The distribution channels that nowadays connect developers in the open hardware movement with manufacturers in the Shenzen region had not yet been established. The design intentionally made use of general-purpose parts that then had to be tweaked to fit the bill. Many hurdles in sourcing the parts arose from the fact that the intended customers were companies rather than individuals or wireless communities. The example of the LED lamps mentioned above is but one case. Another key component was the lens. Commercial FSO devices are equipped with specially crafted optical lenses that can cost thousands of euros. The Ronja device made do with an ordinary magnifying glass. Word had it that a booth in the Prague flea market sold magnifying glasses for a few euros apiece of exactly the right size to fit into the chimney pipes. In neighboring Slovenia, trips were organized to Prague to bring back large quantities of these magnifying glasses and have them distributed locally.

Once the parts had been sourced, the main challenge remained of putting the pieces together into a functioning unit. An experienced builder had to dedicate a couple of days to solder the electronics and carry out error checking. Much of the Q&A on the website and the discussions on the web forum were centered on the electronics. Cutting, drilling, and welding the metal parts following the instructions ran into several weeks of work. A derivative version of Ronja was designed from plastic pipes, originally intended for sewage. It took much less time to customize the plastic parts, although the final product was less robust. This modification was not sanctioned by the official Ronja project, and people learned about the modifications by word of mouth in the Czech wireless community. After the parts had been put together, the equally strenuous task remained of sealing the holes in the chassis with silicon. This was necessary because the Ronja device, mounted on the roof, was exposed to harsh weather conditions. Air humidity could easily damage the electronics inside the hull. It was not uncommon for someone building a Ronja device for the first time to dedicate a couple of months or even half a year to learning everything that was necessary. Testifying to these difficulties is the sight of piles of half-built Ronja devices abandoned in attics and garages.

Once built, the next challenge was to aim the two heads in order to make them connect. To compensate for the weakness of the LED, the light cone had to be tightly focused, and, hence, narrow. The device at the opposite side could be up to a kilometer away. It took an experienced Ronja user a couple of hours to align the two devices. The aiming had to be done in the dark, often on top of a sloping roof. Furthermore, the mounting had to be very steady for the head to remain fixed in position despite strong winds and bad weather. The level of precision in the aiming is suggested by the fact that the link often broke down in the weeks following the mounting of the device, due to temperature changes and compression of the material. Typically, the aiming had to be finely adjusted a couple more times before the connection started to work reliably (Zajicek, December 14, 2008).

The effort, time, and skill that it took to get a Ronja up and running was quite stunning. From the perspective of the philosophy of user-controlled technology, this was "a feature, not a bug." The user had no option but to become knowledgeable about the technology he, or (in a few exceptional cases) she, was using. It contributed to the grander plan of mounting a communications infrastructure that would be transparent all the way down to its users.

RONJA: THE FORERUNNER OF THE OPEN HARDWARE MOVEMENT

The development of the Ronja technology was a collective effort, but with an undisputed leader at its center. The inventor and instigator of the project was Karel "Clock" Kulhavy. A remote control for the family television had given him the idea that he could use light waves to connect his computer with a friend's computer across the street (Hudec, December 8, 2008). It took three years of experimentation before he got the first prototype up and running. He announced the project and invited others to join in 2001. Clock decided to make the design public as a way of justifying the amount of time he had already spent on the project (Kulhavy, November 16, 2008).

It would be another ten years before the global open hardware movement constituted itself as such. Hence, at the time, there existed no dedicated open hardware licenses, no routines for troubleshooting and no distribution networks connecting hobbyists with manufacturers in China. Members of the Ronja community had to discover for the first time on their own the best methods for developing hardware in a distributed and collaborative way. Inspiration came chiefly from the free software movement. It furnished them with software tools, content management systems, and other kinds of supportive infrastructure. Notably, the Ronja design was published under the GNU Documentation License. Along with the software code and the free licenses came the values and political outlook of the free software movement. Its influence is clearly detectable in the notion of "user-controlled technology."

The advocacy of free access to the source code boils down to the recognition that, without such access, someone other than the user will be in control of the technology (Stallman 2002). In addition to the political

analysis, there is a strong emotional and aesthetic investment in free technology. For instance, hackers often speak about the beauty of free software
code in contrast to proprietary "spaghetti" code (Chopra and Dexter 2008).
The novelty of the philosophy of user control lay in the fact that these
sentiments were transposed to hardware development. Subsequently, the
critique of intellectual property extended into a critique of the dominant
market model of developing all kinds of consumer electronics.

The subsequent success of the Ronja project owed much to these values
and ideas. This is suggested by the fact that there were several more scat-
tered attempts in the Czech Republic to connect computers with home-
built FSO links, but without any common sense of purpose attached to
them. Those designs were typically driven by individual curiosity and/
or the need to solve a specific problem encountered by the inventor in
question. One such machine, called Cheapo, had been custom built so
that the inventor, who lived at the bottom of a valley, could connect his
computer to a Wi-Fi antenna at the top of a nearby hill (Seliger, September 21, 2008). What made Ronja stand out was that, from the outset, it
had been conceived with an unknown third user in mind. This anticipated
user imposed many constraints on the design. The parts had to be general
purpose and easily procured. Whenever there was a design choice between
an expensive solution and a labor-intensive solution, the latter prevailed.
Sometimes Clock invested months in tweaking a cheap, general-purpose
component into doing the same thing that could otherwise have been
achieved instantly with a specialized component. No less tedious was the
necessity of documenting every stage of the building process in such a
way that the instructions could be read and understood by a layperson.
On the upside, the public nature of this approach was also what allowed
a community of developers and users to pool their time and resources
into the project. A case in point are the workshops that were organized
in Prague to help first-timers build their own Ronja devices. Volunteers
proofread, illustrated, and made translations of documents relating to the
project. Another tedious task performed by volunteers was to respond to
newbie questions on the mailing list (Sykora, November 27, 2008).

As valuable as these contributions were to the overall diffusion of the
project, they remained at the margins of the actual development process. Part of the superiority claimed for free software development over

commercial programming consists of the capacity to incorporate contributions from disparate sources. There was a lot of tinkering with the
mechanical construction in the extended Ronja community. Improvements in the structure for mounting the head on the rooftop were incorporated into the official version. Another modification to the mechanical
design that became popular, without ever being officially condoned, was
the aforementioned replacement of the metal pipes with plastic pipes.
This was a process innovation for speeding up the construction of Ronja
heads without bringing any advantages to the end product, and Clock
rejected the modification from the official lineage.

A key component of much greater complexity was the electronics used
to translate the incoming light into computer-readable data. In the original
version of Ronja, these electronics had to be soldered together from discrete, air-wired components. It is telling that this construction was referred
to as the "birds' nest." Getting the electronics to work was a major deter-
rent to first-time builders. The option of ordering custom-made printed
circuit boards (PCBs) from a firm was prohibitively expensive at the time.
At first, Clock rejected the many requests to have the discrete, soldered
components in Ronja replaced with PCBs. Some resourceful members of
the extended community took it upon themselves to translate the air-
wire electronics into PCB schematics. Jan Skontorp in Sweden and Ondrej
Tesar made contact over a mailing list and collaborated for six straight
months on the design. In 2003, they made public a working PCB prototype. It spread widely, and Tesar estimates that, for a year or two, their version was the most widely used design modification in newly built Ronja
devices (Tesar, October 5, 2008). Overall, there might have been as many
as four or five different designs of PCBs in circulation at one and the same
time (Seliger, September 21, 2008). Clock was eventually persuaded to
publish an official version of Ronja that incorporated PCBs. On his own
account, he did so reluctantly, in the recognition that it was too tall an
order to communicate in writing to laypersons how to put together the
air-wire electronics (Kulhavy, November 16, 2008).

The long-term viability of the Ronja project was predicated upon its
capacity to stay up to date and be compatible with overall developments in
the wireless network ecosystem. The superiority claimed for the distributed
model of free software development rests precisely upon this capacity. At

least one attempt was made in the extended Ronja community to effect the transition from a 10- to a 100-Mbps connection. Although the job could be done relatively easily with lasers, there were many drawbacks with such a solution from the standpoint of user control. The light cone of a laser being much smaller than that of diodes, it would take a correspondingly greater amount of precision in assembling the parts and then aiming the equipment. In addition to the technical hurdles came the risk that the user would be permanently blinded by accidentally looking into the laser. After much experimentation with LED lamps, two developers succeeded in manipulating them into blinking at the required speed. They nevertheless abandoned the attempt, concluding that the majority of the LED lamps burned out, and there were no indicators to predict beforehand which ones would work. The manufacturer had not accounted for this parameter. Hence, even though they could build an individual device with LED lamps with the sought-after performance, the developers failed to stabilize the equipment or write a protocol that others could have followed. It is highly significant that even in the presence of this failure, the project leader preferred not to make the transition to laser. This suggests that the principle of user control took precedence over the technical performance of the machine. However, as overall development moved on relentlessly, the Ronja device gradually lost its initial edge in functionality over commercial equipment. A window of opportunity was thus opened up for those who did not share the idea of user control, to create a fork in the development project with the goal to maximize the performance of the FSO technology.

THE CZECH WIRELESS NETWORK COMMUNITY

The spread of Ronja took place against the backdrop of the surge in wireless community networks in many urban centers around the world during the early 2000s (Hampton and Gupta 2008). These community networks made use of a small segment of the electromagnetic spectrum that had remained unlicensed by governments, because it was deemed unsuitable for commercial or military purposes. Computer companies started to sell Wi-Fi antennas intended for in-house uses, for instance, to connect computers in office buildings or at trade fairs. As the prices of this equipment dropped, community activists started to build neighborhood networks using the technology (van Oost, Verhaegh, and Oudshoorn 2009; Dunbar-Hester 2009). To put the equipment to such a different use required ingenuity from the activists. It was commonplace, for instance, to build simple antennas from pineapple cans in order to give direction to the signal and make it travel a longer distance (Snajdrvint, December 14, 2008). Mechanical engineering was essential in the construction of wireless networks. Hence, the community activists were among the first to make the leap from software development to tinkering with computer hardware. In the late 1990s, the free software movement was riding a tide of political influence and self-confidence. Its products were conquering the world, most notably the GNU/Linux operating system, together with its peer-to-peer production model for organizing programming tasks (Dafermos 2012). It was not only software tools that wireless network activists culled from the free software movement. In the bargain, they also took on its ideas and values about the importance of protecting political freedoms, sharing information, allowing for collaborative working practices, and so on.

In addition to the influence emanating from the free software movement, the wireless network activists were historically rooted in older forms of independent media production. In larger European cities, the first Wi-Fi nodes were set up in community media centers housed in squatted buildings. The transmission of pirate radio and street TV and the printing of fanzines were directly linked to the broader political outlook of the residents in these settings. The ambition to connect neighborhoods in a locally controlled network grew out of the same activist tradition (Downing 2001; Atton 2004; Juris 2005; Crabu et al. 2015). The prospect of building a communications infrastructure through bottom-up initiatives contrasted favorably with the development trajectory of the internet. In the years immediately before and after the dot-com bubble, the last remnants of the original, end-to-end network were being replaced with the centralized surveillance and advertising machine that we know the internet to be today (Sandvig 2004; Carpentier 2008).

In accordance with the interpretative grid inherited from the squats and the autonomists, wireless network activists typically spoke of the limited frequencies allotted to Wi-Fi transmissions as a form of state censorship. It made the idea of using the light spectrum for transmitting data very attractive, not only in terms of its technical functionality, but also in terms of ideological conviction. In theory, it would be possible to extend the network of optical links to cover a metropolitan area. The Ronja project was heralded, both on its official website and in different discussion forums operated by wireless network communities, as the missing piece in the puzzle in the realization of a decentralized and self-governed communications infrastructure.

The global movement around community networks acquired a local flavor in the Czech Republic due to its members' collective experience of having lived under state socialism only ten years previously. The tradition of samizdat publishing suffused the local activist milieu. A more direct remnant of the communist legacy, however, was the Czech telecoms company. This state-owned monopoly constituted the perfect foe for community activists to rally against. It was held responsible for the poorly developed infrastructure for internet provision compared to other European countries at the time. Slow and expensive internet access was the default standard. Dial-up modems were in use well into the 2000s, which partly explains the exceptional growth of community networks in the Czech Republic. At one point, the country had the highest proportion of citizens in the European Union accessing the internet through wireless technology (European Commission Directorate-General for the Information Society and Media 2009). In Prague alone, there were as many as 250 independent wireless networks. Some of these consisted of just a handful of neighbors in a single street. Other wireless networks counted several hundreds or even thousands of paying members. The largest network among the nonprofits in the Czech Republic was located in the city of Plzen and had more than eight thousand members. There was some coordination of these independent networks through an umbrella organization, the CZFree .net, started in 2002. This milieu furnished the Ronja project with a large user base and technically skilled developers. At the same time, the aims of the two communities diverged, and tensions between them grew. The founders of CZFree.net hatched the idea of developing a replacement for Ronja that would not be hampered by the notion of user control.

USER CONTROL VERSUS DESIGNING FOR MASS PRODUCTION

The amount of expertise, time, and effort that went into building a single Ronja device held back the diffusion of the technology. To someone whose overriding goal was to accelerate the spread of wireless networks, and for whom the FSO link fulfilled a subservient function within this network, the philosophy of user control came to look more and more like an obstacle. The imperative of growing the membership base had both political and commercial undertones. One of the founding members of CZFree.net, who went by the nickname "Deu," operated the first gateway that the many Prague wireless networks used to access the internet. By incorporating the FSO links into this larger network, he hoped to build the demand for his service. He set up a workshop in his basement where people could get assistance in building their Ronja devices. In this basement worked Lada Myslik. He had been experimenting with FSO technology since the 1990s, and he was now charged with the task of helping inexperienced users to find bugs in their Ronja devices. It was a frustrating experience, Myslik recalls, and it convinced him that another design was needed that cut the amateur out of the loop. He envisioned a design where the units could be mass produced and mass marketed. Together with three more companions, Deu and Myslik started a business venture with the ambition of developing such a product. They named their proprietary FSO device "Crusader." The name was borrowed from an old computer game in which space rebels fought against an evil space consortium. These wireless network activists fashioned themselves as the underdogs in an epic struggle against the Czech telecoms monopoly.

From the literature on social movements, examples abound of activists who reframe their political struggle in terms of an opposition to monopoly practices (Hess 2005). With such an interpretative grid, political activism may effortlessly blend into entrepreneurship and Schumpeterian-style "creative destruction." The vision of Myslik and his companions was to sell Crusader to an expanding ecosystem of small internet service providers. Many of these companies had emerged from formerly nonprofit wireless network communities. According to this vision, the old business model in the telecoms sector would be uprooted, leading the way to a decentralized communication network that was resilient against censorship and surveillance. Myslik was nonetheless lucid about the fact that this political vision had already been compromised to some extent by the commercialization of many wireless network community groups: "This was the idea, to have this independent network, which would be immune against eavesdropping and snooping and which would provide people the means of free communication as in freedom. But, as can be seen, people who were more business-like, or, hijack-like, [name omitted], had the idea of going down the business route. And once they managed the network in this manner, it became a business and a telecoms operator subject to Czech laws. If you don't go down this route, it's nothing, it's guerilla" (Myslik, January 9, 2009).

The need to pay for upgrades drove the commercialization of the network infrastructure. In turn, the need for upgrades was driven by increasing expectations about speed and reliability among new members of the wireless network (Polak, January 16, 2009). Related to this development was the growing difficulty of relying on volunteers for maintenance work. In the early days, the wireless networks had been operated by the same group of users who afterward utilized the service. There had not been any shortage of volunteers back then. The situation changed as the wireless networks expanded to involve new users, whose main interest was not to be part of the community and its grander visions, but to have cheap and fast internet access. Maintaining the networks became more of a nuisance, and the surest way to get a task done was to pay someone to do it (Sykora, November 27, 2008).

The commercialization of the wireless network community spilled over into the Ronja community as well. Experienced builders were regularly commissioned to build Ronja links for less knowledgeable users. It became commonplace among the old-timers to have built a handful of devices for a small monetary compensation. It is important to stress that, in keeping with the philosophy of the free software movement, members of the Ronja community broadly approved of people deriving an independent source of income from the technology. The official website directed visitors to small businesses building Ronja devices on demand. When someone asked on the Ronja mailing list if he could sell Ronja devices and keep the profit for himself, he was encouraged to do so (Obadal, Ronja mailing list, November 1, 2004).

It is difficult to estimate how many of these on-off businesses there have been. Experiences from one town in the Czech Republic, Chrudim, suggest that there must have been a large number. Chrudim lies in eastern Bohemia and has roughly twenty-five thousand residents. There were three groups in the town that manufactured Ronja devices for sale independently of each other. One of these groups consisted of four high school students who started operations in 2004. They spent the first year just figuring out how to build links to connect their own houses. In the meantime, they modified the mechanics so that the units could be produced more rapidly. They quickly realized that much time could be saved by replacing the metal construction of the original design with plastic pipes. In total, the group sold ten Ronja links for 550 euros each. Their primary motivation was to have fun, and they wrapped up the business when other hobbies became more enticing (Nemec, December 14, 2008). Shops of the same limited scale emerged in Pardubice, Brno, Prague, and several other Czech cities (Elias, September 10, 2008; Horky, January 17, 2009). It is indicative of how lively this commercial activity was at one point in time that there was even room for some subcontractors within the production chain. A seller in Brno specialized in building the electronics for Ronja, and his customers were mostly other builders who assembled and sold the completed devices (Michnik, December 17, 2008).

Larger sums of money came into circulation when commercially operated internet service providers at the lower end of the market started to use Ronja devices as a backup for their services (Zajicek, December 14, 2008). This enticed the more entrepreneurially minded users to scale up their production process to meet the soaring demand (Horky, January 17, 2009; Michnik, December 17, 2008). One misadventure in Chrudim is highly revealing in this regard. A local businessperson involved the members of the area's wireless community network in a more ambitious business plan. A local internet service provider commissioned them to build the first links. Five employees were engaged at one point in the production process, but it went nowhere. After a number of delays and disappointments, the technically competent partner withdrew from the venture. Pondering the failure, he offered the following observation: "I believe there's something in the design that makes it possible to make it at home for you, but it's not possible to make it for sale." Asked to expand on this line of thought, he dwelled on the command structure going from the businessperson to the employees: "I said that he has employees, but they're doing other projects and I think they felt they couldn't be bothered with Ronja. Because he told somebody who was doing something completely different from

Ronja, I would say, building a house or doing metal boxes, and he told this person to go with me and go on the roof of the building and try to do something. The employee had no motivation to make a success of this project. I guess he had motivation to fail with the project and not be bothered with it anymore" (Kolovratnik, December 14, 2008).

The team behind Crusader had come to the same conclusion. The all-important benchmark when redesigning Crusader was to minimize the amount of manual labor that went into building and mounting the device, as testified by Myslik: "I know at some point I have to get rid of this work. So I give it to some other guy, or to a machine, which is the cheapest. So, I'm aiming for complete machine-controlled manufacture where only this part of aiming is done by people. Even there, I'm preparing for automatic alignment on these units" (January 9, 2009).

With for-profit internet service providers as the intended customers instead of individual users, different design considerations followed. The total cost of the FSO link was considerably higher for a firm since it had to pay at least two employees to spend several hours on a roof aiming the equipment. Myslik quickly realized that the alignment process had to be automated if Crusader was to become competitive. In addition to removing the built-in bias against serial production in the original Ronja design, yet another aspect of commercial product development was the need to keep information secret from competitors. Citing one occasion when a competitor took a design solution, Myslik was thereafter determined to construct his device in such a way that it would be resistant to reverse engineering. This was also the point of direct contestation with the Ronja community, from which the original FSO design had been derived in the first place.

Clock and his followers approved of money being made on Ronja by individual users, but they were insistent that modifications to the design stemming from the common pool of development work must be given back to the community. This policy is consistent with the compromise that the free software movement had previously struck with the profit motive. It amounts to an "information wants to be free" limitation on profit maximization from free software. A peculiarity of writing code, however, is that the development process contains within itself the affordances for tracking derivative versions and, subsequently, to call out violations against the

licensing terms. This possibility of keeping track of derivations is missing in hardware development. The decision about what shall count as
an original piece of work as opposed to a derivative version turns into a
matter of taste and persuasion.

The first version of Crusader looked suspiciously similar to Ronja. It
had the same characteristic double-pipe head, and it used the same LED
lamp. The main difference consisted in the use of PCBs instead of air-wire
electronics. Clock insisted that, since Crusader was a derivation of Ronja,
information about technical improvements must be released, in compli-
ance with the GPL documentation license (Kulhavy, Ronja mailing list, July
20, 2003). The Crusader team refused to make the design public, and
infighting ensued. The schism divided the community as a whole. Eventually, the contention found its way even to the team behind Crusader.
It broke up under the weight of allegations of misspent investments and
money that had gone astray. Afterward, Myslik continued developing a
second version of Crusader on his own. There were several more attempts
in the Czech Republic to conduct commercial development around the
FSO technology in isolation from the wireless network community, but
none fared better than Crusader.

FINANCING TECHNOLOGICAL DEVELOPMENT OUTSIDE
OF THE MARKET RELATION

The debacle over Crusader was also a turning point in the Ronja project.
It demonstrated to Clock that the GPL license lacked enforceability in
the absence of strong community norms. He declared to the Ronja and
wireless network communities that he was done with making his designs
public and up for grabs by others: "I simply refuse to work for free. I've
realized that, after a period of time of working for free, the fridge becomes
empty. And the wallets of various dubious individuals that breach GPL,
lie, use the Ronja project name for direct marketing tricks, and do various
other nasty things on or beyond the verges of the law, become full" (Kulhavy, Ronja mailing list, June 27, 2003; Kulhavy, November 16, 2008).

This kind of bitter experience has been widely reported from other
community-centered development projects. Typically, it is the moment
when developers who initially devoted themselves to the project out of

curiosity or out of idealistic motives decide to start their own firm and try to reap what they consider to be their fair share of the rewards. Clock was determined, however, not to go down that road. He was convinced that the philosophy of user control could not be achieved with a different business model, since this goal was in conflict with the market system as such. He briefly explained why an economic system that mandates that someone is at the bottom and someone else is at the top of the pecking order must result in maliciously designed consumer goods of inferior quality. A major culprit was when revenues were extracted from controls put on the dissemination of the products or, alternatively, on the flow of information relating to the products in question. Some of these ills could be alleviated, he suggested, by exercising income control before the product was released. This was the rationale behind conditioning further developments of Ronja on donations. He hoped that the donation model could make the development process economically sustainable without allowing the project to succumb to market constraints.

The donation model provides a touchstone when reflecting upon the link between technological change and the predominant economic and industrial structures within which most innovations unfold. Arguably, one could choose to interpret the donation model as a micropolitical intervention into the economy. This outlook would be consistent with findings from what David Hess calls technology- and product-oriented social movements, as exemplified by environmental movements developing renewable energy solutions as an integral part of their activism, or patient groups advocating alternative medical therapies (Hess 2005). The roles of social movement actors and entrepreneurial market actors merge under the auspices of the present economy, which consists of a plurality of fundamentally different kinds of markets. The political intervention then consists of the addition of a new product or business model to the preexisting market ecosystem. The donation model of the Ronja project stands out from this analytical grid, however, in that it started from an analysis of the market economy as a single entity, which is to say, "capitalism," which it sought to transcend. The object was to set free time and resources for the development of technology independent of commodity circulation.

Clock explained his donation model on a German mailing list dedicated to discussing the wider political and economic ramifications of free software development. He made a comparison with the General Public License (GPL), the modified copyright agreement used to protect free software code from proprietary claims: "GPL exploits the power of copyright to undermine its own power and create exact opposite—copyleft. I think the Ronja "financing model" could use the power of money to undermine their own power in technology and create the exact opposite—technology, that is free" (Kulhavy, Oekonux mailing list, September 21, 2005).

At the time of this posting, in 2005, the Ronja project had already been supported by donations for about two years. New features that were ready for release were announced together with a price tag. The price was calculated on the basis of time spent and materials consumed in the process of developing the feature. Information was withheld from the public until the requested amount of money had been collected. The information was then disclosed in full at the same time as the product was put into circulation. After the release date, there were no restrictions on the rights of the user.

The most advanced design financed by the donation model was a box called Twister. This was used to transform the incoming light signals into a readable format for the computer. Many in the Czech wireless network community had been longing for the release of Twister. Earlier versions of Ronja had relied on an outdated network card that had become increasingly difficult to find. Clock asked for thirty thousand Czech koruna before he made the design public. The target was met in just six days. During the following two years, three more designs were released under the same conditions.

In spite of this success, the donation model was eventually abandoned. Later improvements, such as the second generation of Twister, were released to the public without any requests for financial compensation. The experiment was successful in that the targets were met, although the sums were relatively small. Some extremely favorable circumstances, that would be difficult to generalize, contributed to this positive outcome: the soaring demand for wireless networks at the time, the tight-knit community supporting the project, and the strong track record of the inventor. Fundraising in the wireless network community provided the bulk of the donations. Many of the community networks, although they were not run for profit, had acquired considerable funds from their large member base by charging subscription fees. For a couple of years, modified Ronja machines ran

on the backbone of these networks. Hence, the strong interest in supporting debugging and the development of new features for Ronja.

At the time when the donation model was introduced, some concerns
were voiced on the Ronja mailing list that development would be derailed
by it. One commentator warned against a vicious circle. If the amount of
money was insufficient, and the development process ground to a halt,
people would become even less willing to make donations (Dalton, Ronja
mailing list, June 30, 2003). Plausible as that scenario sounds, it did not
happen. The main reason for this was that Clock did not try to approxi-
mate the full demand for his product design, in which case the requested
sums would have been much higher. Commentators external to the Ronja
project, taking a more theoretical perspective, objected that the donation
model would reintroduce a quasi-market relation through the back door.
Although formally no sales took place, it was foreseeable that the development process would begin to gravitate toward the demands of the donors-
cum-customers (Merten, Oekonux mailing list, September 27, 2005).

The donation model did not last long enough to confirm or deny this
bleak forecast. Perhaps the confinement of the project to the Czech Republic made the user base too small for this model to be feasible in the long
run. Confirmation of the suspicion can be sought, however, in the commercial platforms for crowdfunding product development, which emerged
at about the same time, and which have since become a key component of
the tech start-up milieu. From just the short period for which the experiment with donations was up and running, we have some indications that
it contributed to widening the gap between the main developer, Clock, and
other advanced users in the Ronja community. Ever since the beginning
of the project, decentralizing the production of FSO units was high on the
agenda, but decentralizing the development of the design itself between
the main developer and other community members received less attention, leading to tensions. Numerous experiments with FSO links were
conducted by members of the Ronja community, feeding the development process with bug reports and requests for new features. However,
few of the design modifications stemming from the extended Ronja community made it into the official release (Bohac, September 14, 2008).
Advanced users whose main interest in the project was to tinker with FSO
voiced frustration now and then on the Ronja forum. They complained

that the discussion threads were overflowing with requests for help with malfunctioning devices built according to instructions. Although the experimental designs rarely made it into the official releases, they sometimes spread anyway by word of mouth within the Czech and Slovakian wireless scene (Sykora, November 27, 2008).

There could be many explanations for this. Making substantial, valuable contributions to the core development of Ronja required expertise in electronics, optics, and mechanics, a combination of skills that greatly reduced the pool of potential collaborators. Even in free software projects, strong power laws are at work behind the ostensibly distributed manner in which the software is being written. In the development of hardware, it is even more of a challenge to divide and allocate tasks. As Clock recalls, people added features without taking into consideration whether the modification would work for others or under different conditions. Alternatively, they did not devote time to writing instructions that could be understood by nonspecialists (Kulhavy, November 16, 2008). It is quite a remarkable thing that an open development community such as Ronja was geared toward meeting the demands of inexperienced users as opposed to being centered on satisfying the curiosity of the most knowledgeable users. That said, the contradictions within the donation model were a contributing factor. The response that Clock gave to a question on the Ronja mailing list is telling in this respect, when he suggested how the community could contribute to the development of the next generation of faster Ronja links: "Yes, in sending financial gifts up to such density that thanks to Ronja I wouldn't have to waste time going to work and could work on Ronja full time instead" (Kulhavy, Ronja mailing list, June 23, 2005).

From this reply, we can see how a division of labor had reasserted itself over the development process. The fact that the requested donations were calculated on the basis of the number of hours that Clock had spent solving a problem strongly suggests that the donation model reintroduced a monetized way of conceptualizing time. This can be contrasted with how another member of the Ronja community reflected upon his involvement in open hardware development projects. Having grown accustomed to seeing people embedding his ideas in proprietary developments, the person in question felt that it was a waste of his time to chase them down. He attributed his lax attitude to not having any ambition to earn a living from hardware development (Simandl, October 27, 2008). The same idea was expressed in a discussion thread on a Czech mailing list comparing Ronja to Crusader. In response to a previous posting that had put the price tag for Ronja much higher than that of Crusader due to the "opportunity costs" of the time spent building and aiming the former, the first commentator retorted, "How much does the time cost for drinking at the pub? / Kolik asi stoji cas propity v hospode?" (CZFree mailing list, November 3, 2003).

No doubt, the unspoken presupposition when reasoning in this way is that the person in question has his or her upkeep covered in some other way, be it through the "family wage" of one's parents or spouse, unemployment benefits, or, as was most common in this case, student allowances/ debt. Indeed, the expansion of the educational system, with the concomitant growth in the student population in many countries during the past few decades, has underwritten the flourishing of many nonmonetized development projects, technological or otherwise. The push to commercialize such activities typically comes at the point in time when the respite in the education system is up. It is to the inventor of Crusader, Myslik, that we owe a statement succinctly capturing the structural constraints under which he, Clock, and all the other hobbyists have to make choices about the course that the technology should take: "You get a job, and you end up working eight hours a day and then you travel quite a large amount of time like to someplace where you live, and then you don't have friends because all those friends are in some other place, and you pay for whatever to let you survive and then you figure out what this world is about. It's about selling your work to those people who can buy it and make profit from you. I didn't want to let anybody make profit on me because that would make me a slave" (Myslik, January 9, 2009).

THE RONJA PROJECT COMES TO AN END

In February 2018, a self-described crypto-anarchist announced on the Ronja discussion forum his intention to update Ronja, in order to deploy FSO technology within the activist milieu. Upon receiving the answer that the project had reached a dead end and been abandoned, the crypto-anarchist responded, "It was a wonderful idea, maybe it came too early before Snowden, and we had not enough maturity to understand the advantages of such project, and the catastrophic Nazi police state we are all in now" (Bonne, Ronja mailing list, February 8, 2018).

To this, an old-time member of the Ronja community responded that interest in the project had dwindled in spite of full anticipation of this bleak future, because broadband internet access had become widespread and cheap. In this short exchange, the fundamental tension is captured between, on the one hand, the political ideas and values of the project, summed up in the label "user-controlled technology," and, on the other hand, the products' functionality, convenience of use, and competitive pricing. The two are not antithetical in any straightforward sense. The success no less than the failure of the user-controlled and surveillance-resistant FSO device hinged on its technical functionality within a competitive milieu. Whereas 10 Mbps was a stunning speed when the project was launched, commercial providers of the same service caught up over the following years. This observation actualizes anew the question of how a hacker project can detach itself from mainstream economic relations and industrial standards while remaining technically efficient and up to date, so that it also has a larger political and societal relevance.

It is worth stressing that, although commercialization eventually delivered the final blow to the project, commercial forces were also conducive to the hackers' political goals. Due to the vast amounts of time, effort, and skill that it took to build and operate a Ronja link, it was rather an exclusive thing to be involved in. Mass production and the black-boxing of consumer electronics lowers the threshold for entry and diversifies the user base. This commonplace notion has a bearing on the Ronja project due to its overriding purpose: to diffuse the open FSO device in order to replace the centralized infrastructure of the internet with a fully decentralized communications network. The wider circulation of the FSO device within a nascent commodity market expanded the user base and opened up some avenues for political action. Concurrently, however, other avenues were closed down, chiefly by turning users into regular customers, whose involvement in the project was conditioned on getting value for money.

We gain a clear indication of when interest in Ronja peaked from the shops selling the device. A vendor in Brno reports having received their last commission in 2007. Most of the orders came during 2005 to 2006 (Michnik, December 17, 2008). The moribund moment is captured in the anecdotal observation of a bin full of discarded Ronjas that had been dumped by a local internet service provider in spring 2008. If we left it at that, we would be reinforcing the standard interpretation of a relentlessly advancing technical frontier that left behind an evolutionary dead end. The motor of this development would be ascribed to a linear growth in technical functionality. However, the demise of the Ronja project has more interesting lessons to teach us than the observation that, at a certain point, the technology reached an evolutionary dead end.

A change in Czech regulations in 2006 to open up 5.5-GHz frequencies for Wi-Fi transmissions was decisive. It is curious to observe that the previous, more restricted regulatory situation, which the community network activists described as a form of state censorship, had provided the background conditions for allowing the open FSO device to flourish in the first place. These legal changes did not have an immediate effect on the wireless networks. This much is suggested by developments in Slovakia, where the same regulatory decision had been made a year before, without Ronja being replaced at an earlier date in the neighboring country. Incompatible standards between Europe and the United States made it difficult at the beginning to construct wireless networks using 5.5 GHz.

The prospect of making the project relevant again by upgrading the FSO link to send data at 100 Mbps was often debated on the Ronja discussion forum. The discussants recognized, however, that even if a technical solution was found, the advantage in speed would not last very long. In those discussions, we get a glimpse of the Herculean difficulties of putting the philosophy of user-controlled technology into practice. The interest of ordinary people in using the technology hinged on the alternative outperforming the industry, not just in the functionality of a single artifact, but in having a development model that sustains innovation throughout. Arguably, this is the extraordinary accomplishment of the free software movement. The distributed model for developing source code in the open is as much of a moral and ideological feat as it is a logistic and technical operation.

It is noteworthy that, even one or two years before the industry standard for wireless communication had caught up with the Ronja project, the collaborative development process had wound down due to infighting among its members. The decline began with the schism over conflicting entitlement claims. When there was no market value in the product, the notion of financial remuneration never arose in anyone's mind. The reward for individual contributions to the open development process consisted of higher status among one's peers, whose opinions counted for something as long as the community thrived. With the anticipation that profits could be made on the technology, further improvements in the design started to look like competitive advantages and trade secrets. Conflicts flared up on the Ronja discussion forum when someone claimed to have made a break-through but then refused to give any details about their discovery. Those who were looking for collaborators to work on some aspect of the tech-nology in order to release the results to the public complained that their calls were met with silence on the mailing list. It was rumored that some-one had discovered a way of making the LED lamps send signals twice as fast as normal, but this could not be confirmed. The development of Ronja came to a standstill because the energy was channeled into propri-etary spin-offs instead. The developer of one such invention, a 100-Mbps optical device provisionally named "Cyclop," estimated that there might be three or four more development projects in the Czech Republic at the same time. It tells us something about the state of the community that neither he nor anyone else could determine the number of such projects that were underway (Kamenicky, December 4, 2008).

From the above discussion, we conclude that the dwindling perfor-mance of the open FSO technology relative to commercial development played an important role in the story of the demise of the Ronja project, but not exclusively so. Moral and ideological support in the community for freely disclosing information had broken down prior to the loss of the technology's competitive edge. A comment from one member of the Ronja community highlights the importance of ideology construction in deter-mining this outcome. While claiming to be strongly committed to the free software movement, citing the threat of state surveillance, privacy concerns, and so on, he did not think that defending the same freedoms in the sphere of hardware development was a moral obligation to the same extent. He offered this opinion as an explanation for the proprietary spin-offs of Ronja. Clock's authority to enforce the GPL license was limited in the absence of full backing from the community (Zajicek, December 14, 2008).

On the whole, the leadership style of Clock resembled that of Rich-ard Stallman, the founder of the free software movement. In addition to calling out violations against the GPL license, Clock upheld an ideo-logical line about the kind of improvements that were accepted into the official release. One member voiced his mild disapproval with the party line by describing the project as "over-Clocked." According to this per-spective, the purist stance resulted in fragmentation rather than compli-ance. From a different perspective, the same thing could be described as a failure to narrate the philosophy of user-controlled technology and the stakes involved. The philosophy mandated trade-offs in the technology's functionality and user convenience, as happened, for instance, when the option of upgrading the device to incorporate a laser was turned down, or in the refusal to adopt designs streamlined for serial production. Many found this too high a price to pay for ideas that they did not believe in to start with. The original design was sidestepped by forks that sought to optimize functionality and reduce production costs, instead of being oriented toward making the technology transparent "all the way down." This underlines the claim by the self-described crypto-anarchist that the premature demise of the project was due in part to a missing sense of political urgency at the time when the project was flourishing.

CONCLUSION

We have told the story of Ronja to illustrate what a case study of the full life cycle of a single hacker project might look like. It serves our didac-tic purposes well that this particular hacker project was delimited in time and space. It lasted from 2001 until roughly 2007 and was geographically confined to the Czech Republic and neighboring countries in central and eastern Europe. The Ronja project foreshadowed the rise of the global movement around open hardware development that emerged over the following years. The Czech project stands out, however, in being guided by the philosophy of user-controlled technology, and in its experimen-tation with a donation model that partially sustained the development process without direct market transactions. This makes the Ronja project exceptional, not only in relation to the mainstream, hierarchical mode of developing technology within a corporation, but also in comparison to the alternative, i.e., the entrepreneurial culture that engulfs most open hard-ware development projects today. The exceptionality of Ronja throws into relief the extent to which mass production, global consumer markets, and waged or entrepreneurial means of earning a living make up the framing conditions for innovation processes in general.

It is evident that a technology developed inside a corporation will be entirely shaped by management hierarchies and investors' prerogatives (Rowland 2005; Croissant and Smith-Doerr 2008; Renee 2017). Indeed, the bureaucratic form of exercising control over technology development is so overpowering that it conceals the more subtle ways in which design choices are nudged into complying with the goal of profit maximization. The necessity of making a living by selling one's labor is the overarching framing condition for technology development everywhere. It goes a long way toward explaining why innovations that emerge outside of formal and institutional settings nevertheless end up reproducing the same priorities and constraints as those found inside formal institutions. Some design choices simply make more sense once the decision has been made to streamline a product for mass production and mass consumption.

Rising market demand for FSO technology in the Czech Republic stimulated an overhaul of the original technology in line with the requirements of mass production. Several attempts were made within the extended Ronja and wireless network community to develop proprietary forks from the technology that had been released to the public. The violation of the GPL documentation license constitutes a textbook example of how recuperation processes work by enclosing the information commons. As is typically the case, infighting ensued. Competing claims of individual entitlements to the common design divided the community, and improvements in the design were no longer shared. The common development project dwindled, but so did the commercial spin-offs. The recuperation attempt failed in that no product was successfully brought to the market based on Ronja technology.

One interesting and unexpected outcome of the schism was Clock's experiments with financing development through donations. He did so out of the conviction that the principle of user-controlled technology was irreconcilable with the fundamentals of a market economy. His aspiration was to sustain invention processes independently of the mass market and the patent system. It should not come as a surprise that this attempt failed in many respects. The donation model could not disengage from the totality of market relations, and, subsequently, it was rife with contradictions. That being said, several updates of Ronja were made under this model, suggesting the dedication and cohesiveness of the larger community from which the project drew support.

What we referred to in the theory chapter as the three pillars of autonomy were mostly in place in the Ronja project. Technical skills were widely diffused in the extended wireless network community, allowing for informed judgments to be made on issues such as the political ramifications of a centralized as opposed to a distributed communications network. Victories scored and defeats endured by an earlier generation of free software developers provided the informational backdrop for the Ronja project and were regularly invoked by Clock and by discussants in the dedicated online forum. As suggested by the donation model, the project was supported by a community with a shared value system, although this crumbled as tensions rose in connection with the commercial derivatives.

This suggests to us that the fate of an individual hacker project such as Ronja cannot be assessed in isolation from the overarching, second and third, time horizons within which the project's life cycle unfolds. In relation to the second time horizon, the overall development trend within hacker movements, the Ronja project is situated at a point in time when the influence of the free software movement had begun to wane. Over the following years came a reorientation within hacker culture toward commercial and instrumental goal fulfillment. In terms of the third time horizon, capitalism as an evolving whole, the flourishing and eventual demise of the Ronja project were intimately tied up with developments in the world market and policy making at national and EU levels. On the one hand, the dissemination of the Ronja device relied on a global market in consumer electronics, from which it sourced, among other things, generally available, off-the-shelf parts. On the other hand, the conditions for the Ronja project to thrive had inadvertently been created by Czech regulators, who sought to protect the state-owned telecoms monopoly from global competition. When the import restrictions on Wi-Fi equipment were lifted and more frequencies were opened up for transmitting data (decisions that were taken at the behest of EU lawgivers), user demand for the FSO device was pulled from under the feet of the wireless community.

4

OPEN-SOURCE 3D PRINTING
REPRODUCING MACHINES
AND SOCIAL RELATIONS

Most people we sell to are ready to live in the future, who want to own the
means of production and feel empowered to make things themselves.
—Bre Pettis, entrepreneur, December 6, 2010

The above statement was made by a start-up entrepreneur selling desktop
3D printers under the brand "MakerBot." The quote is full of Marxist bra-
vado, giving a hint of the utopian air that surrounded 3D printing technology during the early 2010s. To the entrepreneur, no doubt, this rhetoric
served a marketing purpose. However, the fact that the scenario sounded
plausible to the ears of a mass audience is noteworthy in itself. In the business press and official policy documents, personal manufacturing and mass
customization were hailed as a panacea for the many ills that had been
caused by previous waves of industrialization: alienated factory work, the uni-
formity of mass consumer items, the outsourcing of jobs from the shores
of North America, and wasteful supply chains spanning the whole planet
(*The Economist* 2011). After desktop 3D printing hit the peak of the "Gart-
ner hype cycle" in 2012, the utopian hopes that had once been invested in
the technology in many different quarters of society, not least in academia,
started to look quaint. Ten years ago, the hyperbole ought to have been
subjected to relentless critique. Now, in contrast, the time has come to take
those forsaken utopian claims seriously.

Both the technology of desktop 3D printing and the associated utopian ideas originated in a community of open-source hardware hackers. The stated goal of these hackers was to democratize general-purpose manufacturing capacities in much the same way as the members of the legendary Homebrew Computer Club had done for computing in the 1970s. Analogous to how the original hardware hackers envisioned a computer that could "run on the kitchen table," the spread of desktop 3D printers was expected to put an end to the dominant logic of centralized "mainframe" manufacturing. This outcome was guaranteed by the design of the 3D printer itself, such that the machine could make a copy of itself. Accordingly, the name of the project was "RepRap," an abbreviation for "self-REPlicating RApid Prototyper."

Alongside the launch of the development project in 2004, the project's instigator, Adrian Bowyer, published a manifesto in which he outlined the economic and political ramifications of having unleashed a self-replicating, ubiquitous manufacturing unit onto society. It was from this manifesto that the entrepreneur quoted above borrowed the rhetoric about reclaiming the means of production. It is against the benchmark defined in Bowyer's manifesto that we talk about RepRap as a concluded project, in spite of the continued popularity and market penetration of desktop 3D printing. We concur with how one longtime contributor to the development of RepRap assessed the project's legacy: "RepRap was greatly successful on one of its major goals, making affordable 3D printers available to everyone. It wasn't too successful on the other goal, triggering an evolution of self-replicating printers" (Hitter, email communication, August 21, 2020).

A closer examination of the history of the RepRap project is warranted by the fact that the open-source desktop 3D printer serves as a stepping-stone for numerous, ongoing open hardware projects. It was foundational in the launch of the open hardware movement, only matched in importance by the open-source microcontroller "Arduino" (Mota 2014). Referring back to the terminology we introduced in the theory chapter, recuperation within the time horizon of this individual project has sedimented the landscape within which the open hardware movement is currently unfolding. Moreover, the RepRap project was unique among open hardware projects in that it came attached with an explicit ideological program; another such rare case is the Ronja project, with its philosophy of user-controlled technology. The manifesto provides us with an entry point for investigating the unstated political imaginary of the open hardware movement in general.

The scenario of democratizing manufacturing capacity that was articulated in relation to fused deposition modeling, the technical name for 3D printing, is rehearsed in hacker projects dedicated to a range of other manufacturing techniques: open laser cutters, open lathes, open CNC mills, and so on. Concurrently, the open-source machine park is about to merge with cloud computing and blockchain technology. What is bubbling up within the open hardware movement is nothing short of a reconceptualization of the assembly line from the ground up. It is in this context that we recall the origins of computer-aided machine tools in the history of industrial conflicts and union busting. A handful of budding experiments in the RepRap community with cloud manufacturing give additional support to our central claim in this chapter: rather than a democratization of production, the diffusion of manufacturing capacity across society is just as likely to result in an extension of "factory despotism" into the private sphere. Although all of this might sound overly far-fetched, history advises us to pay close attention to engineering utopias. When engineers dream about a brighter future, their dreams tend to come true, albeit in an inverted, nightmarish form.

DESKTOP 3D PRINTING PAVING THE WAY
FOR DARWINIAN MARXISM

The basic idea of 3D printing is that physical objects are made by extruding a material (usually plastic) in multiple layers. It offers a highly versatile manufacturing technique. In the industry, fused deposition modeling has been used to make prototypes for half a century. It was the termination of key patents on the technology that enabled the hacker project to get under way in 2004. In the hands of hackers, the technology was geared toward making 3D printing affordable and fitted to home use. In the early days, it was a challenge to get the development project off the ground, as it was exceedingly difficult to find the parts to build new 3D printers. Although the plastic material was cheap, it was expensive to have the special parts custom made by firms. Facing this problem, it made sense to design the desktop 3D printer in such a way that it could make some of its own parts. As put by one of the developers, who at the time was a student of Adrian Bowyer, working in his laboratory at the University of Bath, "Having a machine that can print itself, by definition, the cost of it must reduce drastically, because, pretty much no-one can make any profit on it. And by having it open source means it is cheap to develop (Jones, November 26, 2009).

The technical concept of a 3D printer that can print (some of) its own parts was a pragmatic engineering solution to the immediate problem of acquiring the custom-shaped parts. However, that is not the whole story. This engineering goal was tied to a more utopian vision about setting off a chain reaction that in the long run would render markets in desktop 3D printers obsolete. The price of such machines would fall to the marginal cost of the materials, thanks to the possibility of making copies of the hardware. What is more, the same thing would happen to markets in every other material good that could be "printed" on the same printable 3D printer. If a 3D printer was made so that it could print a copy of itself, it followed that the machine was sufficiently advanced to print just about any other conceivable consumer electronic good. In short, the spread of this ubiquitous manufacturing unit to the masses was hailed as a road map for abolishing the market economy. This idea was concisely expressed in the original byline of the RepRap project, "wealth without money."

The road map for bringing about this economic state was codified in the manifesto alluded to above, with the suggestive name "Darwinian Marxism." The manifesto stipulated that industrial manufacturing, global distribution networks, and mass consumption would be disrupted once the diffusion of the desktop 3D printer had reached a certain threshold. The crux was how to scale up the dissemination process sufficiently to match the production capacities of the incumbent industrial system, without reverting back to mass production. Self-reproduction was offered as the master key to unlocking this problem. The idea was imprinted in the name given to the project, "RepRap." In the same spirit, the first few generations of printers were named after famous biologists. The first version was released in March 2007 under the name "Darwin." It was replaced with a second generation that was called "Mendel." Another, "Huxley," also saw the light of the day, before the number of models multiplied beyond anyone's oversight, and the uniform naming practice was abandoned. In 2008, the project leader and one of the developers in the core team, Vik Olliver, announced that the open-source desktop 3D printer had had a child. Which is to say that, for the first time in history, there existed a second 3D printer that had been assembled from parts printed on the first machine.

This bold statement provoked heated debates within the extended open hardware movement. It was quickly pointed out that only half of the parts of the "child" had actually been printed on the parent machine. This observation invited a more general challenge to the technical feasibility of the long-term engineering goal of the RepRap projects. Skeptics objected that the key components of the machine, such as the electronics, the motors, and the extruder, could never be created through a process of adding layers of melted plastic. Even if some technical hurdles can be overcome in the future, the proposition of self-reproduction runs aground on the logical problem of what could be termed "bad recursiveness." That is to say, the more advanced capacities were built into the machine, the harder it became to print that machine. The idea of "bad recursiveness" can most quickly be illustrated with the proposition of printing the extruder head: the extruder head must be built from a material that can withstand the temperature needed to melt and extrude the material in question.

The technical impossibility of realizing the vision of self-reproduction did not discourage the advocates of the RepRap project. At least discursively, they resolved the challenge by extending the narrative about a life-like diffusion of the machine. The analogy with living beings was still valid, they argued, because biological life forms are also not completely self-sufficient. Vitamins have to be brought in from outside for the organism to sustain herself. The engineering target was set at "bringing down the vitamin count" of the machine: that is to say, to make an ever-greater proportion of the components printable. They responded in the same manner to another common objection, namely that the machine could not assemble itself. In analogy to the wasp and the orchid, the 3D printer reproduced itself in symbiosis with human beings. The human being will assist in the reproduction process of the 3D printer because of the useful goods that they can procure from the machine.

Upon hearing this response, one might protest that, with such a generous definition of self-reproduction, even a regular machine shop would qualify. The factory machinery "evolves" in symbiosis with the machine operator. Although the argument was not put in those terms, the implicit assumption behind the advocates' reasoning pivoted on the idea of functional autonomy. The reason why the symbiosis of "worker–factory" did not fully qualify was because the former has been subjugated as a mere appendix of the latter. The symbiosis "user–3D printer," in contrast, was envisioned to be mutually beneficial, expanding the freedom of action of both parties. Looked at from this angle, the many objections that could be raised on technical grounds were irrelevant. However, the tropes about evolution dressed up a possibly more contentious agenda. The functional autonomy of the community vis-à-vis for-profit entities hinged upon the technical concept of self-reproduction. Thus, the hacker community reassured itself that it was going to remain in charge of events when the inevitable approach from industry came.

The term "self-reproduction" was often used as a shorthand for extolling modular design choices in the project, thereby ensuring that the required parts could be printed or sourced from standardized, off-the-shelf vendors. The benefit of this lay in the fact that no single vendor could lock in the key resources for building the machine. As long as the design was kept open, hackers would be guaranteed the right and capacity to route around any resource limitations or constraints imposed through the design. This tactical reading of the situation was foregrounded by Vik Olliver, the same hobbyist who, together with Adrian Bowyer, had announced the "child" machine to the world: "When people try to make money, more specifically when they try to put something in the way so that you have to go through them to do something interesting, the project generally tends to fall apart. But that doesn't happen with RepRap because it is specifically designed to reproduce itself. So you couldn't really put yourself in the way and demand money" (Olliver, May 4, 2010).

The influence of the free software model is clearly detectable in this quote. The political values of the free software model are encompassed in and realized through the development of programs designed to be modular and transparent. This ensures the user of the software the right to freely use, share, examine, and modify source code. Tied to these four rights of the user is a much broader, political worldview. Proprietary software and closed-design solutions are identified as root causes of monopoly power.

Free software is put forward as a countervailing force against this tendency of information and power to concentrate in the hands of a few. This part of the ideology surrounding free software was successfully carried over to the subsequent wave of open hardware development. Much the same analysis was now applied to the centralized manufacturing of physical goods. A case in point is the following quote from Ed Sells, former student of Adrian Bowyer and principal architect behind the second-generation "Mendel" 3D printers, which until this day remain the backbone of most RepRap derivatives:

I think that Adrian [Bowyer] has hit on a mechanism which is so unbelievably powerful. When you've got something making itself, it is scary from the point of view of HP [Hewlett-Packard]. Self-reproduction wins over anything else, over any linear production. RepRap exposes the fact that if you've got a 3D printer, it can make itself. So HP will go, "Well, we are not going to make any money here," and the fact that Adrian has made it open source from day one means that there is nothing to stop people designing around someone coming in. I don't think you can stop RepRap except if you get to a safe distance and nuke it. (Sells, May 7, 2010)

Ed Sells was referring to an argument in the Darwinian Marxian manifesto that sought to prove the superior capacities of the decentralized manufacturing model of the hacker community over the centralized manufacturing model of industry. Mathematically speaking, printing 3D printers at home had the potential to outperform the mass production of 3D printers in factories. Provided that the question of exhaustible resources was bracketed, the self-reproduction of 3D printers would numerically overtake mass production techniques, as exemplified by an injection molding machine. This would happen with the same force as exponential growth surpasses linear growth.

More important than the brute, numerical advantage, however, was the appeal to the superior dynamics of an open development process. The claim mirrors the famous catchphrase of open-source software guru Eric Raymond: "Given enough eyeballs, all bugs are shallow." The basic message is that innovation accelerates more rapidly the more people are involved in it. Consequently, an open and decentralized development process (of software) will win out in a race against a closed and centralized development process. This article of faith of open-source advocacy combines with tropes taken from evolutionary biology to reinforce the belief that the hacker community will always have an edge over firms selling nonreproducible 3D printers. This point is clearly stated by Joseph Prusa, whose fame in the open hardware community stems from having designed the spin-off "Prusa" 3D printer, which has since become the de facto standard: "We are primarily the community, and the community accesses evolution: bad ideas which do not work die off, good ideas go ahead. We have that. But if you have like a company at the center of the community and the community around it, . . . the company has the force to overcome this evolution, so they can choose what to do. In RepRap you can't. You can do whatever you want but if it isn't good it won't work" (Prusa, September 19, 2011).

It is significant that the above three quotes come from Vik Olliver, Ed Sells, and Josef Prusa, arguably the three most centrally placed developers in the RepRap project after its founder, Adrian Bowyer. In other words, the utopian aspirations invested in the development of an open-source, ubiquitous manufacturing unit were not the ravings of a few mavericks or "hacktivists" coming in from the sidelines. The dream of bypassing the monetary system by creating "wealth without money" was a key motivational force driving the development process forward. Just as important, however, is the timing of the above quotes. They were uttered at a key juncture in the project's history, a year or two after the first start-up firms had been launched. The pull from rapidly growing consumer demand for desktop 3D printers, corresponding with the anticipation of future business opportunities in the sector, had just started to make its presence felt in the community. The utopian claims served to boost confidence and a sense of purpose among the project's key developers in the face of upcoming challenges.

Although the aspiration of unleashing economic and social disruption upon incumbents was attested to by some of the developers in the core team, it is fair to say that this ranked low on the priority list of most of the engineers involved in the RepRap project. Their motives for being involved varied greatly. As has been observed in other hobby projects, the simple joy of tinkering with technology is a major factor in motivating people to take part in collaborative development (Kleif and Faulkner 2003). Other motives were the possibility of getting a home-built desktop 3D printer at a cut-rate price, being part of the buzz that surrounded this technology for a couple of years, and, for quite a large number of participants, latching onto the anticipated business opportunities. However, the possibility of gathering all these viewpoints and motivations under one and the same umbrella was itself part and parcel of what made the call for (biological) diversity so appealing to some of the participants. The meanings invested in the term "diversity" were analogous to the notion of "pluralism" in activist milieus. It signaled the antidote to party lines, sectarianism, and divisive politics.

This brings our discussion to an inherent tension within the Darwinian Marxist program that influenced the course the RepRap project was to take from that moment onward. The many references to Darwin in the manifesto were embraced, or at least endured, by the participants on the RepRaP discussion forum. In contrast, objections were often raised to the mentioning of "Karl Marx." The name of the revolutionary was perceived to be in conflict with the value of diversity (RussNelson, RepRap forum, August 27, 2007). With a nod to Herbert Marcuse's critique of the idea of tolerance, the political outlook of the RepRap community could be called "repressive diversity." As the technology gained more traction, the Darwinian Marxist manifesto was downplayed in the self-presentation of the project. The philosophical and political debates about the possible disruptive consequences of the general-purpose constructor, which had once animated the discussion forum, dwindled away. It is telling that the byline of the project, "wealth without money," which had originally been displayed at the top of the front page, was removed from the website during a design update in 2010.

The marginalization of utopian voices in the developers' community coincided with the maturation of the consumer market in 3D printers. Contrary to what one might have expected, however, talk about an imminent revolution due to the dissemination of desktop manufacturing continued as before. For a time, this revolutionary rhetoric was carried on by the start-up firms (an example of which was given at the outset of this chapter) in a bid to grab public attention, market shares, and potential investors. Intriguingly, no one expressed alarm over orphaned phrases from Marx's oeuvre when they were uttered by corporate mouthpieces. This suggests to us that the "real thing" was easily distinguishable from marketing hype by those who were in the fray. As we will see below, the emotional investment of the RepRap project in the notion of diversity played a decisive role in how the community regulated (or not) its inter-actions with the for-profit interests.

PUTTING A NEW SPIN ON ENGINEERING IDEOLOGY

As eccentric as the program of "Darwinian Marxism" might sound, the history of the engineering profession suggests that its claims and imagery drew on a utopian tradition stretching back to the nineteenth century. A cornerstone of the imaginary of the engineering profession is that human emancipation will march forward hand in hand with the advancement of science and technology. The marriage between reformist politics, tech-nological progress, and appeals to the innate laws of nature can be traced back to Saint Simon, a militant during the French Revolution. Many of his followers were engineers who had been educated at École Polytech-nique, a school that was established by the Jacobins in the aftermath of the revolution. Already during the *ancien régime*, engineers had begun to discern a dynamism in nature which they vaunted as a model of effi-ciency. This idea of nature carried a political payload, because dynamic nature was contrasted with the blockages and inefficiencies of the feudal order of the day (Picon 2009). In the updated version of this narrative, as retold by the RepRap developers, technological development is blocked by incumbent monopolies, intellectual property law, management pre-rogatives, and centralized modes of manufacturing goods.

The propensity among engineers to anchor their ethical and political claims in nature was given new impetus with the breakthrough of evo-lutionary biology in the nineteenth century. The name to be mentioned here is not Charles Darwin, however, but Herbert Spencer, who was an engineer himself (Sharlin 1976). Spencer's notion of social Darwinism served as a founding ideology for the engineering profession in the late nineteenth century. Although never developed into a single, coherent doc-trine, some core ideas continued to resurface in the engineering profession with great consistency over the next century. A key assumption was that nature and society are governed by laws that are immutable and univer-sally valid, yet accessible to human comprehension and manipulation.

In the social sciences, assumptions about deterministic laws of nature are typically considered to be a limitation on the agency of human beings

(Wyatt 2008). This is not the case for engineers, however. As Francis Bacon put it: in order to command nature, one must first obey her. It is by obeying-commanding nature's laws that engineers throughout history have made their bid to extend their profession's influence over society and over other social groups. One must resist the first impulse to decry this attitude as an expression of abusive privilege and informational advantage. We would then have to rest content with a half-truth. Historians of technology have argued that the popularity of these ideas surged at a time when the subordination of the engineering profession under corporate bureaucracy was being consolidated. In the face of adversarial developments, an assurance of professional values and identity could be sought in this narrative about engineering nature's laws (Layton 1986). In fact, the same line of argument is applicable to the discredited technological determinism of the early workers' movement. It reached its highest pitch at the moment when the Second International was being torn apart by belligerent nationalism (Söderberg 2011). Antonio Gramsci's balanced reflection on the determinism of the early workers' movement surpasses most things that have been said on the topic by constructivist scholars. He warned that determinism leads to "passive and idiotic self-sufficiency" in a social movement. Worse still, it encourages deference among the rank and file toward the party leadership. In the same breath, however, he granted that deterministic conviction had given fortitude during times of setbacks (Gramsci 1971, 646). We will return to this observation below to make a key point about the RepRap project.

Engineering ideology is a double-edged sword. An illustration of this is the ambivalent relation of white-collar engineers to the labor question. Their professional identity and value system were extended and codified under Taylorism. Frederick Taylor and his followers believed that they had discovered immutable laws about management that had the same validity as nature's laws. They imagined the engineer to be an impartial judge of those laws. Thus, the engineer lifted himself above the messy world of politics, and especially the antagonism that raged between labor and capital. Paradoxically, it was this antipolitical outlook of the engineer that made them, in their own eyes, so well suited to meddle in political affairs and industrial conflicts. Taylorism combined the promise of enhancing the efficiency of industrial production, benefiting the whole of humanity, with a bid to enlarge the sphere of influence of the engineering profession over the shop floor. No doubt, this would be at the expense of blue-collar workers and trade unions. That being said, engineers, starting with Taylor himself, portrayed the scientific approach to management as a pushback against the autocratic prerogatives of top managers of the old days (Layton 1986, 139). Scientific management was given an explicitly anti-big-business twist by one of Taylor's associates, Morris Cooke, who urged his colleagues to guard public interests against vested interests and tycoons.

The most radical exponent of this outlook was Thorsten Veblen, who expressed his ideas in the pamphlet *The Engineers and the Price System* (2001). Although his agitation had only a marginal influence on later developments in the engineering profession, the grudges he voiced against the irrationality of letting the profit motive dictate industrial development were shared by many rank-and-file engineers at the time. Veblen charged the business community with obstructing the forces of progress and depriving the public of the full blessings of science and technology. In an industrial society, he asserted, engineers were the ones most qualified to run the show, not the captains of industry. He called for a "soviet of engineers" to take charge of America's productive base (Veblen 2001).

As we know, the engineering profession did not heed this call. An important contributing factor was, no doubt, that the personal fortunes of engineers were so closely tied up with those of the industrial magnates. Revolving doors between engineering assignments and the upper echelons of management weakened the ability of engineering societies to assert their autonomy vis-à-vis business. Of no less significance is the fact that, from an early date, the business community was determined to take control of the curriculum of engineering schools. Through the educational system, engineers were made to conform to the status quo even as they unleashed wave after wave of disruptive technological change (Noble 1977). At the end of the day, it was not market forces, Layton persuasively argues, but bureaucracy that aroused the strongest resentment among engineers.

This historical lesson resonates with the surge of software programming as a recent addition to the engineering profession. As we noted in the theory chapter, the mindset of the computer engineer was decisively shaped by the 1960s counterculture. An unsympathetic interpreter of this history, Alan Liu, asserts that the main achievement of scientific management was not the subjugation of blue-collar workers under capital. It was

the creation of a new stratum of workers with a persona perfectly in keeping with scientific management. This product of Taylorism merged with
its opposite, the countercultural "bad attitude," to make up the strange
amalgam that he calls "cyberpolitics" (Liu 2004). From this grew the Californian brand of management philosophy (Barbrook and Cameron 1996).
This brief digression into cyberpolitics gives an idea of how Cooke's and
Veblen's indictment of free-market forces as antithetical to engineering
principles and human rationality could so easily be transformed into an
opposition to market regulation and bureaucracy. The demand for a democratization of the means of production merges with an idealized picture of
a free-market cottage industry. Framed in this way, the critique takes aim
at the institutions that ensure a semblance of employment security and
the tempering of market arbitrage.

To remove market demand for the living labor of other professions has
always been the bread and butter of the engineer's job. Once upon a time,
however, this task was undertaken with a word of regret and possibly a
half-baked promise of more rewarding employment opportunities to be
created in the future. Now, however, the same task has attained a radical, messianic fervor with the culture of politicized computer engineers.
To prove the point, recall the debate over filesharing that raged during the
early 2000s. Activists in the filesharing movement were unapologetic about
the fact that professional musicians had to learn to swim or sink under the
torrent of technological change and fickle consumer demand (Andersson
2011). This message was taken to heart by youth activists in the Swedish
Pirate Party. These activists visited furniture fairs to pass on the message to
IKEA salesmen and designers that their jobs were soon to be extinct thanks
to the coming wave of disruption set off by desktop 3D printers. The vendors of design objects would end up on the same garbage heap of history
as musicians and record company managers (Nipe, December 23, 2009).
This anticipated outcome was also welcomed on the RepRap discussion
forum under the heading "democratization of the design profession."

It might be considered a redemptive feature of this discourse that the
hobby engineers did not exempt themselves from the "democratizing"
force of factory automation. Indeed, the collective identity of hackers is in
itself a testimony to the crisis of the engineering profession. The historian
of technology and former dean of MIT, Rosalind Williams, is well placed
to reflect upon this crisis. From the ever-more evanescent engineering

curriculum she sees a loss of identity of the profession as a whole. She attributes this in part to the tendency of engineering practices to merge with scientific research into something that often goes under the name "technoscience." Equally significant, however, is the disappearance of the institutional settings within which lifelong engineering careers used to unfold. Granted, precarious labor demand is a condition known to workers in many other branches, but in Williams's opinion, engineering students distinguish themselves by having fully internalized the entrepreneurial outlook. She laments this trend, because it hollows out the public commitments that were integral to the professional identity of the engineer in the days of Veblen and Cooke (Williams 2003).

Williams gives an insightful account of the ongoing transformations within the engineering profession. Her divination about the profession's future is somewhat gainsaid by the statements in the Darwinian Marxist Manifesto. The text reveals strong thematic continuities in engineering thought and a sustained willingness to appeal to the public interest, in spite of the fact that those ideas are now being articulated from extremely precarious positions outside of the contractual employment relation. It is evident that hobby engineers are prone to being captured by an entrepreneurial opportunity structure and the corresponding free-market discourses. It is debatable, however, whether hobby engineers are more vulnerable in this respect than were the employed engineers of the twentieth century, who became integrated into the management hierarchy. A common theme in both cases is the eulogizing of automation. Whereas this eulogy had a ring of complacency about it when voiced by the employed engineering profession, the same cannot be said when this ideology is acclaimed by those who have themselves been deprofessionalized and pushed into precariousness due to automation. Hackers, due to the position from which they speak (i.e., outside of the salaried profession) put a different spin on the old discourse about automation as a vector of universal human emancipation.

AUTOMATION AS PERIL AND PROMISE

The present debate in the news media and academic publishing about the rise of "algorithmic" or "platform" capitalism (Bratton 2015, chap. 9;

Srnicek 2016) is just the latest iteration in a long series of periodic debates about the threat of automation leading to a growth in joblessness and rising inequalities. A milestone in this debate was Harry Braverman's book, *Labor and Monopoly Capital* (1974). Its status as a modern classic can be discerned from the storm of criticism that was leveled against him and the school of labor process theory that he helped to instigate. The two bones of contention were whether or not deskilling should be treated as an inherent tendency of capitalism (Wood 1987) and the extent to which managerial rule was secured through winning the consent of the workers, as opposed to coercing them into compliance (Friedman 1977).

One empirical field was infallibly called upon to test these claims, namely the application of computers to machine tools. The seminal work that must be mentioned here is David Noble's study of the introduction of computer numerical control (CNC) machinery in the manufacturing industry. He documented how CNC machinery had been embraced by scientists, policymakers, and business leaders as a solution to the problem of militant trade unionism. The CNC machine held out the promise of doing away with the know-how, and consequently the effective control, that blue-collar machine operators exercised at the shop floor level. By programming the movements of the machine tools in advance, managers hoped that this knowledge could be positioned with the white-collar computer engineers instead, who were considered to be more trustworthy. Ultimately, the goal professed by the promoters of this technology was to have a fully automated factory without any workers at all. Noble discussed the many ways in which this ambition was frustrated. Cooperation from the machine operators, at least during the initial phase, turned out to be necessary in order for the CNC machinery to run properly. Once their cooperation had been secured through various concessions, the CNC technology could be perfected to the point where it could eventually be rolled out as it had first been conceived (Noble 1977).

At the time when Braverman's and Noble's books were being debated, the jury was still out over the long-term consequences of factory automation. Almost half a century later, we may safely conclude that industrial trade unions have been much weakened. Although this outcome stems from multiple causes, no explanation would be complete without mentioning the impact of computerized and automated manufacturing techniques.

A telling anecdote is the encounter that one of the authors of this book had with a spokesperson for the US metal workers' trade union in the 3D printing pavilion at a Maker Faire in New York. Surrounded by the latest descendants of CNC machinery, the spokesperson handed out stickers urging the visitors to buy products made in the United States. Trade union politics has been decimated to the point where employment security is now no more than a consumer preference. The industrial conflicts that culminated in the dissolution of the Fordist composition of the working class (i.e., blue collar, unionized, class conscious) has left a "material trace" in the form of the desktop 3D printer.

From this vantage point, we may formulate a response to the critique of Braverman and Noble voiced at the time. Their critics objected that computerization would not eradicate skill in an absolute sense, it would just relocate the demand for skilled labor. New tasks and occupations would replace the lost ones. Computer technology had concurrently given rise to an expanding workforce of software programmers (Senker and Beesley 1986). Indeed, managers started to complain about programmers as a nest of labor unrest as early as the 1960s: "The technologists more closely identified with the digital computer have been the most arrogant in their willful disregard of the nature of the manager's job. These technicians have clothed themselves in the garb of the arcane wherever they could do so, thus alienating those whom they would serve" ("The Thoughtless Information Technologist" 1966, 21).

Likewise, the introduction of high-level programming languages has regularly been hailed in the business press as the solution to managers' dependency on skilled programmers. This line of reasoning is exactly the same as what was heard back in the days when CNC machines were introduced into the manufacturing industry (Ensmenger and Aspray 2002).

The analytical categories that Braverman drew upon were sufficiently dynamic to also be applicable to the freelance programmer. In the opening statement of his book, Braverman made a plea for an open-ended definition of the working class, so that the analyst could follow the cat-and-mouse game between capital and labor through its ever-shifting manifestations. At one point in the argument, he made the observation that, when workers sought refuge from the boredom of a semiautomated workplace in nonremunerated, spare-time activities, capital followed closely in their trail: "So enterprising is capital that even where the effort is made by one or another section of the population to find a way to nature, sport, or art through personal activity and amateur or 'underground' innovation, these activities are rapidly incorporated into the market so far as is possible" (Braverman 1974, 193).

This observation justifies our analytical strategy of studying the application of CNC-like machinery in contexts outside of the contractual employment relation with reference to labor process theory. It points us, in other words, toward hackers engaging in the "amateur or underground innovation" of open machine tools. Not only are desktop 3D printers descended from past industrial conflicts, but so is the identity of the nonemployed hardware hacker. The defeat of the Fordist class composition, as documented by Braverman and Noble and others, has disrupted the transmission of identity constructions and collective memory from one generation to the next. This link has been completely severed in the passing from the factory to the FabLab, in spite of the fact that they are connected by numerous "material traces," starting with the machine tools that many FabLabs have inherited from defunct manufacturing companies in their proximity. Consequently, during these new cycles of struggle, hackers and makers have to invent a language of their own to articulate their grievances. The intellectual resources that have been handed down to them come from the winners of previous rounds of struggle over automation: that is to say, from managers and white-collar engineers. The fragmented experiences of class antagonism are thus articulated as a longing for the total automation of society.

The narrative about a self-reproducing, ubiquitous manufacturing unit acquires a new, oppositional meaning when this narrative is expressed from a position outside of the salaried engineering profession. Inadvertently, the hardware hackers have picked up the utopian stream in engineering thinking that dates back to Saint Simone and Thorsten Veblen. More surprising, perhaps, is the fact that the potential for emancipation in automation was not lost on Braverman either. Toward the end of *Labor and Monopoly Capital*, the author acknowledges that technological development contains the potential for abolishing the division of labor: "The re-unified process in which the execution of all the steps is built into the working mechanism of a single machine would seem now to render it suitable for a collective of associated producers, none of whom need spend all of their lives at any single function and all of whom can participate in the engineering, design, improvement, repair and operation of these ever more productive machines" (Braverman 1974, 320).

FROM REPRAP UNION TO MAKERBOT INDUSTRIES

The hardware hackers in the RepRap project aimed to disrupt the existing industrial mode of mass production by disseminating desktop 3D printers to the masses under an open license. This ambitious goal was put to the test by the problem of bootstrapping the project into existence in the first place. The question was how to get the first machines into the hands of developers and so expand the community. The answer, in keeping with the professed goals of the project, was to use the existing stock of desktop 3D printers to make more such machines. For this purpose, the RepRap Research Foundation was launched as a nonprofit institution. It was entrusted with the task of keeping tabs on the latest components and distributing them for the cost of materials and shipping. After the announcement in 2008 of the "birth" of the first RepRap "child," a machine park was set up in the mechanical engineering department of the University of Bath, consisting of four 3D printers dedicated to printing more parts for new machines.

The idea was that the operation would scale up exponentially as hobbyists around the world started to replicate the undertaking. People who had received parts from the factory in Bath were encouraged to produce a second batch of parts and give it away for free to someone else in the community. The required community norms were promoted on the dedicated RepRap website and forum. Occasional postings on the forum from people announcing that they had spare parts to give away suggest that the model worked up to a point. There was a handful of attempts to render the dissemination process more systematic. One proposal, going by the name "RepRap Union," was to recruit local RepRap user groups and hackerspaces as relays in a globally spawning distribution network for 3D printed parts.

The beating heart of the decentralized and self-evolving system of automation was a moral economy. The engineers sought to reinvent a Kula exchange ring of printed plastic parts. The moral code sustaining the original

Kula ring, as famously described by the anthropologist Marcel Mauss in *The Gift* (Mauss 2016), was the expectation that the recipient of a gift would pass it on to someone else in the ring. Through such gift exchanges, relations of mutual gratitude and trust were woven among members of the tribe, summoning the group into existence in the process.

The experiment of creating a gift exchange ring in printed plastic parts came to a premature end. It ran into the same chicken-and-egg problem that held back the diffusion of the machine in the first place, as suggested by a lucid remark by one of the core developers: "Everyone prints out a set of parts and passes them on. But that has not worked because nobody received any parts in the first place. For example, I made my own using that machine [pointing at a RepStrap]. Or people bought them from places like Bits from Bytes and MakerBot. So they don't feel any obligation to start printing free parts for people. If someone received a set of free parts, then they have a moral obligation to print some parts and pass them on. But hardly anybody received any parts so that process has not started up really" (Palmer, March 17, 2010).

In the absence of strong community norms to back up the would-be moral economy, the void was filled with market incentives. As news about the self-replicating 3D printer spread across the world, thanks to intense media coverage of the topic during the early 2010s, demand for such a machine soared. The output capacity of the self-organized, hobbyist production line fell far behind market demand, and from this arose a short-lived, speculative bubble in selling plastic parts. The transition from a gift-based to a market-based mode of circulation was catalyzed by the experiences of numerous members of the community of having been taken advantage of. One participant declared on the RepRap forum that he had stopped giving away printed parts after he had found one of his sets being offered on eBay for two hundred dollars (Spacexula, RepRap forum, July 21, 2011).

In just a short time, eBay became the central distribution mechanism for 3D printed parts within the RepRap community. In addition to resolving the problem of distribution, this fostered a culture of microentrepreneurship in the 3D printer community. Almost all of the members of the core development team had at one point or another sold printed parts on a one-off basis. This revenue stream could be systematized by dedicating a couple of printers to the production of parts for sale. The practice was generally welcomed as a service rendered to the larger community. The growth of the community of potential developers hinged on the successful diffusion of parts for building more RepRap machines.

The next logical step was to scale up commercial operations within a legally recognized entity. The first garage firm producing desktop 3D printers was Bits from Bytes. It had been started in 2007 by one of the members of the core team, Ian Adkins. The founder and his business partner had formal engineering backgrounds and they needed to catch up with the ethos of openness in the community to which they were catering (Adkins and Major, November 26, 2009). Complaints that Bits from Bytes fell short of its obligations in some respects—among other things, that it kept its firmware close to its chest—were voiced on peripheral blogs run by individual developers (Higgs, November 3, 2011). On the whole, however, the firm kept a low profile and avoided drawing negative attention to itself. In 2010, Bits from Bytes was acquired by a global manufacturing company, 3D Systems, for an undisclosed sum of money. Transformed into a subsidiary arm of this multinational, Bits from Bytes disappeared onto the sidelines.

This contrasts sharply with the controversy surrounding the second garage firm that was seeded from the RepRap project, MakerBot Industries. In just a short time, this firm ended up as the bogeyman of the entire open hardware movement. MakerBot Industries was started by Zach "Hoeken" Smith, manager of the RepRap Research Foundation, in 2009. Together with two business partners, he set up headquarters in a hackerspace in Brooklyn, the New York Resistor. The street cool of the open-source firm was skillfully exploited by Bre Pettis, one of the other business partners, as exemplified by the quote given at the outset of this chapter. Within the RepRap developers' community, however, the news was received with cautious skepticism. The founding of the firm coincided with a drop in the stock of supplies of the nonprofit foundation. The foundation's website quickly fell into disarray and shortly afterward it was officially disbanded. The concerns expressed on the discussion forum—prophetic in the light of what happened later—was that the RepRap project, in the absence of a neutral governance body that could act as an honest broker, would be sidetracked by special interests (Ppeetteerr, RepRap forum, March 17, 2009).

Confirmation of this suspicion can be gained from a core developer who resigned from the RepRap project during its early days, Forest Higgs. He recalls that, in the same meeting during which he and the other core team members heard about Hoeken's start-up for the first time, project leader Adrian Bowyer concurrently informed them about his private investment in the new venture. This disclosure, in Higgs' opinion, preempted the possibility of upholding a cohesive policy toward firms, especially in regard to the enforcement of the GPL agreement (Higgs, November 3, 2011). Adrian Bowyer started his own company in 2012, RepRap Pro, as did many other members of the core team over the subsequent years.

It bears mention that the firms selling lightly modified, rebranded versions of the RepRap 3D printer made an important contribution to the dissemination of the technology and, subsequently, to the expansion of the community (de Bruijn, November 11, 2009). Customers were spared having to navigate through a labyrinth of outdated blogs and forum posts in order to find the parts that were up to date and compatible with one another. As the threshold for building the machine was lowered, more people (and people from a greater variety of backgrounds) became involved. Thanks to the commercial kits, the machine park of 3D printers grew exponentially. This expansion owed a great deal, however, to centralized mass production. The goals professed in the Darwinian Marxism manifesto had at some point along the road been tacitly abandoned.

When stray participants in the RepRap community protested against perceived breaches of the license agreement, they gained little traction with fellow hobbyists. Discussion threads containing such allegations were moved by the system administrator from the general forum to less frequented subforums. Tellingly, indignant commentators in those discussion threads did not target the alleged violators of the GPL license, but instead vented their anger against the person who had called attention to the wrongdoing. The latter were accused of offending the value of diversity by expressing hostility toward free enterprise. This suggests to us that norms for enforcing the license agreement and "governing the commons" were not only lacking in the RepRap community. Under the auspices of "diversity," a countervailing norm of *enrichissez-vous* had taken root. Darwin was turned against Marx.

Another indication of this is the mixed responses to MakerBot Industries' fateful decision in 2012 to backtrack on its former open-source policy.

It had been the most brazen voice among the corporate actors, endlessly peddling variations on the "sharing is caring" slogan. Indeed, its product line had been derived on the back of the RepRap community's developers. When the news came that the next generation of MakerBot 3D printers was going to be shipped with closed source, it was widely received in the open hardware movement as a betrayal. Among the most vocal critics was Zach "Hoeken" Smith, who by then had been ousted from the company. Today it is *comme il faut* to decry MakerBot Industries as a manifestation of corporate evil, so much so that some distinctions tend to be lost in the process. In fact, the pariah status of the MakerBot brand name was not firmly established until two years later, when it became publicly known that the firm had covertly patented designs uploaded by "makers" to its open-source repository, Thingiverse. At the moment when Bre Pettis announced the new policy, he could still count on sympathy from many quarters in the open hardware movement. The decision was seen as a regrettable but understandable reaction to numerous, hostile copycat actions by competitors, an experience shared by many self-employed makers and vendors in the open hardware movement.

Of particular interest to us is how the announcement was received in the RepRap community and realigned at short notice with the overall metanarrative about self-reproducing machines. Many called for Adrian Bowyer to speak out against the firm's move to close the source. He was generally looked upon as the moral leader of the community, and, being the founder of the RepRap project, he was the one who had primarily been wronged by MakerBot Industries. He responded in the following way: "If you are taking part in the RepRap project, then I hope that you believe Open Source to be a morally and politically good thing, as I do. But if you don't believe that, you are still welcome to take part, by me at least. When it comes to the success or failure of RepRap, moral beliefs, legal constraints and the flow of money are almost completely irrelevant. It is the evolutionary game theory that matters" (Bowyer, MakerBot blog, September 21, 2012).

In the same posting, Bowyer disclosed to his readers that he owned shares in MakerBot Industries. One would thus not have been surprised to see an uproar against his quip about the irrelevance of intentional human action. There was nothing of the sort. Bowyer's intervention was enthusiastically greeted by commentators in the discussion forum. Elevated to the status of clairvoyant, his response began to circulate on numerous other blogs and forums in the extended open hardware movement. The myth about unstoppable evolutionary forces played the part that Gramsci had once assigned to determinism in the workers' movement. That is to say, belief in such forces exacted deference from the rank-and-file members toward their party leaders.

REPRODUCING MACHINES OR REPRODUCING SOCIAL RELATIONS

The founding myth of the RepRap project contained two storylines that seemingly contradict one another. One storyline professed the law-bound, quasi-biological dissemination and proliferation of self-reproducing machines. The other consisted of an astute analysis of the political implications of design choices, urging engineers to get involved in the action. The contradiction between these two messages is less staggering when seen in the context of the open hardware movement's ongoing struggle to assert its functional autonomy. The hardware hackers assumed that the disruption they sought to unleash with their self-replicating, ubiquitous manufacturing unit would only come about if the open-source development model won out against commercial product development in terms of efficiency. This was widely spoken of as an evolutionary arms race. Being disadvantaged from the outset, the hardware hackers sought reassurance in the notion that the open development process would speed up the evolution of RepRap 3D printers compared to closed and proprietary product development, thus giving the community an edge over the firms.

This tenet of faith was attested to in the distinction made between "RepRap" and "RepStrap" machines. Before the first "child" of RepRap had been delivered to the world, hobbyists made custom parts for 3D printers on home-built, one-off contraptions constructed from metal, bamboo, or Meccano. These machines were referred to as "RepStraps," coined from the word "bootstrapping." RepStraps could produce parts for RepRap 3D printers, but they could not make copies of themselves. As had already been outlined in the Darwinian Marxist manifesto, firms would be compelled by market forces to produce RepStraps, since they had no incentive to sell machines that could make copies of themselves. That being said, the commercial 3D printers could be used to make new RepRap machines. The one-directionality in this diffusion of machines underpinned the argument that community-designed RepRap machines would prevail over commercial 3D printers (Bowyer, November 24, 2009).

The sincerity with which this conviction was held by members of the RepRap community is suggested by the strenuous efforts and oceans of time that some of them dedicated to designing parts for RepRap so that they could be printed on commercial-line desktop 3D printers. Design files stored on the now-defunct blogs of individual RepRap developers and in half-forgotten 3D repositories still bear witness to this undertaking. An example will have to suffice for the sake of illustration. At a point in time when MakerBot Industries was the number one open hardware firm, and its products, in the eyes of the general public, were synonymous with "desktop 3D printing," the company released a new design that, although it complied with the GPL license, was defective in terms of self-reproduction. The print area of the commercial machine was too small to print the parts used in the most advanced RepRap 3D printer at the time, called Mendel. This prompted a quick response from developers in the RepRap community. The mechanical construction of the Mendel 3D printer was redesigned so that it could be built from smaller parts that fitted the print area of the commercial 3D printer (Sells, May 7, 2010).

With the benefit of hindsight, it must be acknowledged that this design intervention, along with innumerable others of the same kind, did not do much to sway the course of events. The concerted action was nevertheless important as part of a community-building ritual, whereby individual members confirmed their allegiance to the common goal of making a self-reproducing, open, and modular 3D printer. Upon this engineering goal hinged the functional autonomy of the community. The last point can be underscored by contrasting this vision with its antithesis, a commercially controlled, closed-source 3D printer shipped with cartridges. It was widely recognized in the RepRap community at the time that this must be the logical endpoint if the 3D printing industry had its way, a forecast that has since been confirmed. Markets in ink printers had already demonstrated that profitability does not stem from selling the machines, but from selling the refills. Maximization of profits required a vendor lock-in of the sales of the cartridges. For this to happen, the architecture of the desktop 3D printer had to be walled off in order to prevent customers from tinkering with the critical component, the deposition head, through which the melted filament passes. This is analogous to the legal and technical control that software firms assert over proprietary source code. Advocacy of an open, modular, and copyable (i.e., self-reproducing) material architecture of the 3D printer was a cornerstone of the recursive politics of the RepRap community.

In practice, this outlook was complicated by the fact that the technical capacity of a 3D printer to print a second copy of itself is not a straightforward either/or proposition. Many different engineering considerations impinge on the design process: the cost and availability of the components used, what tools are required to make those parts, the tolerance range of the components, constraints on available design choices from the overall mechanical construction—the list goes on and on. Some notable issues that were endlessly debated inside the core development team concerned the choice between less or more advanced motors (DC motors versus stepper motors), the kind of electronics to use (single- or multilayered circuit boards), and whether or not to include expensive, hard-to-get components in the mechanical structure of the printer (such as ball bearings). At a second-order level, opinions diverged as to whether or not it was acceptable to use proprietary software tools when designing the 3D parts (Sells, May 7, 2010).

To get a flavor of these discussions and how committed some developers were to the concept of self-reproduction, we focus on the issue of bearings. Bearings are used to facilitate motion. They serve a critical function on the axes along which the printhead moves. The first release of RepRap, called Darwin, made no use of bearings. When the component was introduced in the second-generation Mendel printers, it caused a small uproar. Bearings are expensive and hard to get hold of outside Europe and the United States. From the perspective of self-reproduction, the trade-off looked as follows: On the one hand, bearings increased the accuracy and quality of the prints, which would be passed down to the derivatives. On the other hand, the chances of one day being able to print a complete 3D printer were reduced when bearings and other advanced components were incorporated into the mechanical design. The gains in print quality were such, however, that the use of bearings won general approval. But even then, a handful of developers paid tribute to the principle of simplicity by developing an alternative and simplified version of Mendel without bearings (Olliver, May 4, 2010).

This anecdote about the bearings was brought up in several of the interviews with core developers as a showcase of how potential conflicts among themselves were resolved in the project. The project leader, Adrian Bowyer, refrained from taking top-down and potentially contentious decisions on the overall direction of the project. There was no need to do so, it was generally believed, since those solutions that proved to be the best would gain traction. If someone objected to a particular design choice, the right course of action was to invent something different and better (Jones, November 26, 2009). The notions of diversity and evolution provided the justification for the hands-off approach to project management.

A tacit presupposition of this line of reasoning, however, was that the engineering goal of self-reproduction would remain the objective of the evolutionary selection process. Evolution was called upon to work out the best means for achieving this end. What really happened, however, was that the selective pressure transformed the ends just as effectively as it did the means. As we saw before, just about any choice among the available trade-offs in the design of a 3D printer could be discursively aligned with the idea of self-reproduction. Concurrently, however, some objectives, such as accuracy and printing speed, were in high demand aside from whatever contributions they made to the fulfillment of the original engineering goal. With the speculative bubble in printed parts, new kinds of incentives emerged in the community, tilting the development process as a whole toward mass production and consumer demand. This was attested to by a core developer (aka Nophead) in his response to the question of whether commercialization held back some aspect of development: "Yes, I think the majority of people wanting a 3D printer want something cheap, easy to build and operate with good print quality and care little about it being self-replicating, so naturally there aren't many people working in that direction" (Palmer, cited in Hodgson 2013).

This comment exposes a flaw in the reasoning that the aggregation of spontaneous design choices would automatically lead to a self-reproducing universal constructor. Someone must rig the game, starting with the choice of publishing the results under a free license, and then keep the game rigged—for instance, by sanctioning violations of the license and community norms, for the right kind of evolution to unfold. We arrive at the same conclusion from listening to the two business partners behind Bits from Bytes. They made it clear from the outset that their aim was to develop a commercially viable "RepStrap" 3D printer. Earlier in the same interview, the founders recalled that they had teamed up and started the firm after having met when picking up their children from kindergarten. When asked to reflect upon the overall direction of the development process, they drew a comparison between their personal life situation and the situation of members of the RepRap community: "If you look at the people who are involved in the core team, who are driving it forward, most of them are, I would say, semiretired or they don't have children. So they are people that have the time and energy to devote to it that they don't need a return on" (Adkins and Major, November 26, 2009).

Differently put, in order for there to be self-reproducing machines, it will take engineers who do not reproduce themselves. If this tongue-in-cheek statement is read in a more theoretical register, not implying the "engineer" as a child-rearing individual, but the engineer as a social relation, then the remark is spot on.

A LABOR PERSPECTIVE ON THE CLOUD FACTORY

The utopian aspirations originally professed by the RepRap project, to enable wealth without money with the help of a self-reproducing, ubiquitous manufacturing unit, was abandoned at the halfway point. Still, the desktop 3D printing technology that matured in the process serves as a building block in a line of other hacker projects that continue to thrive. The vision of unleashing political and economic disruption by distributing access to manufacturing capacities continues to lure the open hardware movement into concerted action, although without the elaborate philosophical and political underpinnings of a manifesto. Consequently, the promise that the dissemination of machine tools will bring about a democratization of the "means of production" lacks support in a sustained analysis of the balance of forces in the world. These Marxist concepts are frequently courted in the pronouncements of the open hardware movement without the implications of the terminology being given much thought. The labor perspective is missing, as can be glimpsed from an early experiment in the RepRap community with moving the production of printed parts for 3D printers "into the cloud." Contained in this embryonic form of what was then enthusiastically talked about as the "cloud factory" are the first traces of an emerging division of labor outside of the contractual employment relation, separating the community of developers from a cloud of piece-rate machine operators.

The focus of the RepRap project was on integrating general-purpose production capacities into a single machine. Fused deposition modeling is highly versatile, but it comes with some built-in drawbacks. The principle of constructing an object by adding layer upon layer of material makes this production technique unsuitable for scaled-up and rationalized production. From the outset, 3D printing was suitable for prototyping but not for industrial mass production. To achieve the goal of ubiquitous manufacturing capacities would require a combination of production techniques: CNC mills, lathes, laser cutters, robotic arms, and so on. Indeed, there are numerous, ongoing open hardware projects aiming to bring all of these machine tools into the commons. Furthermore, the tool head is but a single point in the complete manufacturing cycle, which must also include energy provision, waste disposal and/or recycling of raw materials, several layers of embedded control software, and repositories of 3D designs. These auxiliary functionalities of the manufacturing process are also being developed under the auspices of open hardware licenses. Taken together, a material infrastructure for commons-based peer production communities is under construction.

The individual projects to develop one or another aspect of the manufacturing process are overshadowed by even more ambitious plans to synthesize these many functionalities into an integrated, automated, and flexible manufacturing system. This synthetic vision has been formulated in a project called "Cubespawn." The goal of the project is to build a system that can convert a file into a complex physical object under autonomous, automatic control. To do so, the different machine tools are placed in standardized and modular 600-mm aluminum T-slot framed cubes. At the time of writing, the finished parts of the system consist of a RepRap-derivative 3D printer (Ultimaker), a CNC machine, and a robotic arm (Jones, August

11, 2020). However, the real action is in the development of standards and software to connect the individual machine components. With Cubespawn, the assembly line is being reconceptualized from the ground up.

The next logical step is to distribute the assembly line to a network of workstations. For this purpose, a digital platform called Makerverse is under construction to sit on top of Cubespawn. Although Makerverse is still in an early stage of development at the time of writing, the comprehensiveness of the undertaking is breathtaking and warrants closer inspection. The proposal is to connect the manufacturing layer to an online environment for robotic engineering simulation. The design process can then be gamified and integrated into a version control system. Drawing on blockchain technology, the Cubespawn/Makerverse system is envisioned to enable functionalities that are indispensable in mainstream industry, such as verifying that the production process complies with established engineering specifications and industry certifications (Mockridge, August 17, 2020). Admittedly, these ideas are coming from the margins, but no more so than the prospect of an open-source desktop 3D printer was in 2004.

Indicative of the more general trend in the open hardware movement is the mushrooming of digital platforms, for the most part run by companies on a for-profit basis, that propose to integrate distributed, physical manufacturing capacities with layers of control software and networking capacities. A case in point is Wikifactory, which has been up and running since 2018. Catering to open hardware projects as well as businesses, it offers a range of tools and repositories for collaborative product design, all integrated into a single workspace/platform.

In fact, the gravitational pull of the cloud had already made its presence felt when the RepRap project was just about to take off. The capacity of the 3D printer to print some of its own parts was leveled by self-employed vendors of printed parts and by some of the start-up firms. Corresponding to this was the need for remote control of the workflow of fleets of 3D printers. One stab at catering to this demand was an open-source software project called BotQueue, started by Zach Hoeken after he had been pushed out from MakerBot Industries. The experiments with cloud manufacturing in the RepRap community have lessons in store with a bearing on how the landscape of open hardware projects is likely to pan out over the coming years.

It took only a year or two for the revenue streams that many hobbyists derived from selling printed parts for 3D printers on eBay to fall away. A factor contributing to their diminishing returns was the open innovation model itself, because it mandates free disclosure of process innovations. Thus, there was a continuous, downward pressure on the price of the printed parts. Arguably, this is confirmation of the prediction that the self-reproducing 3D printer will undercut markets for all kinds of consumer goods that could alternatively be printed on the machine. Alas, the first people to be affected by the price squeeze were the self-employed hobbyists. Consumer demand for the firms' branded products continued to soar for several more years. Differently put, the firms consolidated their grip over the consumer end of the 3D printing market at the same time as the prices for printed parts dropped on eBay. It is known from the market in open-source software that the ban on information secrecy creates a strong first-mover advantage. The actor with the largest capacity to rationalize production, which is just a different way of saying "concentration of capital," stands to win the lion's share of an open, collaborative development process. This was how the gospel of "sharing" was supposed to work to the advantage of the desktop 3D printing firms. Eventually, however, they too were undercut and put out of business by even larger manufacturers, who were even better placed to take advantage of the copycat logic.

Facing steadily shrinking margins, the self-employed hobbyists started to bicker among themselves over the technical and aesthetic merits of some of the process innovations that had begun to proliferate. One such innovation that was intensively debated was the introduction of cast parts to replace printed parts. This innovation originated in the mechanical engineering department of the University of Washington. Casting the parts made it easier for the university to provide students with their own sets of parts for building a 3D printer. By using a mold, thirteen sets could be cast in the same time that it took to make a single set of printed parts. The next step in the automation of RepRap production came from inside the RepRap community: a master for making the modules to be used to make the parts. Both the molds and the master were very quickly put up for sale on eBay. Coming full circle, these masters were redesigned so that they could be printed on a RepRap 3D printer. The predictable outcome of this frenzied drive toward automation was that cast parts were being sold at prices close to the cost of the plastic material. Not everyone expressed joy about seeing this implication of wealth without money coming to fruition.

In the commentary section of the blog hosted by the University of Washington, grievances about the price squeeze were voiced by Josef Prusa. Today, he is one of the very few commercial vendors of RepRap 3D printers who has managed to stay afloat at the consumer end of the 3D printing market. At the time, however, he voiced complaints that echoed a trade union position in the old debates over automation: "Thing is, that selling parts kinda supports the further development. Someone sells parts, makes some money on that, and he can invest them back. Now, with these printable masters, and I must say they look nice, there will be piles of 'fotons' making moulded parts, which will take the price down even more, and it will slow down the development" (Prusa, Open 3DP, February 15, 2011).

His complaint was rebuked by other commentators, who charged Prusa with standing in the way of the forces of progress. In keeping with the long tradition of engineering thinking (Layton 1986), falling prices were seen as a neutral gauge of technological efficiency and human progress. The heated exchange over the merits of cast parts is rich with insights into the ideological tensions within the open hardware movement in general. The praise for unfettered automation sits uncomfortably with the other major trope of the movement, the figure of the independent and self-employed developer, to whom is assigned the role of a Jefferson's yeoman in the information cottage industry. A more plausible scenario for what it will be like to toil in the "cloud factory" can be derived from the work tasks distributed on corporate-controlled, digital platforms, notably Amazon Mechanical Turk (Irani 2015a).

Indeed, a stab at imitating the crowd-sourced work model in the manufacturing of physical goods was attempted by MakerBot Industries. The firm wanted to outsource parts of its in-house production line to former customers. In the early days, MakerBot kits were assembled in a redbrick, former factory building in Brooklyn, which also housed the hackerspace New York Resistor. The kits were assembled manually in a way that was reminiscent of a nineteenth-century workshop. It is no surprise, then, that the firm struggled to keep up with soaring consumer demand. One bottleneck in the production line was four pulleys. In 2009, three interns worked just to make this critical component. When the interns went back to school after the summer break, the company had to find a replace-ment. MakerBot Industries made the following announcement on their blog: "If you've got a 3D printer, you can print 608 pulleys for the Mak-erBot, and MakerBot Industries will buy them from you for a buck each, with a minimum batch size of thirty for obvious reasons of scale. In time, MakerBot wants to move more of their manufacturing off their factory floor and into the cloud. Future versions of the MakerBot will have first their pulleys, and eventually a hefty fraction of their parts made by other users." The announcement went on to draw out the implications of the initiative: "This is the beginning of a new system of manufacturing, pos-sibly every bit as important as the Industrial Revolution. As time goes on, it won't just be MakerBots that are made by distributed manufacturing, it'll be many things" (Pettis, Thingiverse blog, August 11, 2009).

The firm offered one dollar for each pulley. The community of "mak-ers" produced several thousand pulleys in this way, enough to furnish five hundred to six hundred printers. Besides producing the items for MakerBot Industries, the contracted makers also contributed with process innovations. For instance, instead of producing the pulleys one by one, the printed object was redesigned so that the items could be printed seven at a time and then snapped off from a tray. Although the experiment in cloud production resolved the bottleneck in the production process, it did not scale well. The pulleys had to be produced to meet certain tolerances, giv-ing rise to the problem of enforcing quality control. This much can be read from another posting by the company, this time with a more acid tone of voice: "When we make them, the bearing press fits into the pulley and yours should too! Don't forget to check the pulley for bearing fit before sending them off, because we certainly will!" (Hoeken, MakerBot blog, August 6, 2009).

At the time when this initiative was taken, issues of quality could be settled by phoning the person in question. As the volumes continued to grow, from producing twenty machines a month to a couple of hundred, it was no longer possible for the managers to maintain a personal relation-ship with the "makers." The lack of means for scaling up quality control was, according to Bre Pettis, the reason for ending the experiment (Pettis, September 20, 2011). The difficulty of controlling the quality of the out-put from a remote location could just as well be described as a difficulty in asserting managerial control over a geographically dispersed workforce. Ten years down the road, the company released MakerBot CloudPrint, software for controlling prints from a centralized, digital platform. Thus, the technical infrastructure is emerging for asserting "quality control" in the cloud factory. The dream of the cloud factory is about to be realized in the form of a digitally mediated putting-out system.

MakerBot Industries' prematurely aborted experiment with subcontracting piece-rate work tasks to its former customers underlines the peril faced by commons-based peer production communities, due to the rise of cloud computing. By controlling the graphical user interface, providers of cloud services have introduced a new layer for extracting value from users, regulating access, and consolidating market power over the commons, while concurrently respecting the letter of the free and open licenses (De Filippi and Vieira 2014). Likewise, the emancipatory promise of the open hardware movement (i.e., its contribution to the democratization of the means of production) is likely to go into reverse gear when all the steps in the "open" design and production chain are integrated into a single, multipurpose, and company-controlled digital platform. In the absence of a countervailing ideological program and community sanctions, the "spontaneous" inventions in the open hardware movement are likely to evolve in the same direction. Design efforts will be pulled by market demand and pushed by legal threats from rights holders. There will be financial rewards for those individuals who are so positioned that they can channel the collaborative efforts of the open hardware community into providing technical solutions to the problem of asserting managerial control over a decentralized manufacturing (labor) process.

As a direct outcome of these efforts, the material infrastructure is put in place for an ever-finer division of labor outside of the contractual employment relationship. Engineers, for the most part self-organized into communities of peer producers, sit at the top of the value chain. Their task within the emerging social division of labor is to furnish capital with novel design concepts and experimental development work. The task of beta testing, ranging from the more technical end of the spectrum to pure market research, is outsourced to crowds of users. At the bottom of it all are a variety of menial tasks relating to desktop manufacturing, performed under digitally enforced supervision by a dispersed cloud of click workers.

CONCLUSION

We have offered RepRap as a case study of a hacker project that failed to resist a recuperation attempt. The outcome of this failure was innovation. No doubt, to most observers, including the vast majority of the former RepRap developers, bringing a product innovation to market scores as a success. Reflecting back on the project, Vik Olliver affirms that he and the others in the core development team left after their efforts had been crowned with a fully operational 3D printer (August 12, 2020). Without questioning this insider account of past events, we cannot help but note that the cited accomplishment falls far short of the stated goal of the project at its outset: to unleash a machine upon society that would create wealth without money and thus render the need for markets in consumer goods superfluous. Drawing on interview material that has been collected over a period of more than ten years, supplemented with archived material and numerous searches in the Wayback Machine, we believe we have demonstrated that this utopian idea was widely and sincerely held by many developers in the core team. It served as a key motivational force in the development process. Adhering to the method of "immanent critique," we have sought to reconstruct the distance that has been traveled between the goals that were professed by the hobbyists at the beginning and what those goals are said to have been when the project came to an end. Within this discrepancy, we find grounds for claiming that the project fell to a recuperation attempt.

The emergence of a market in consumer-grade desktop 3D printers provides the ultimate vindication of the claim that recuperation has taken place. Decisive in this story is a series of predatory moves by entrepreneurial members of the community to enclose the common pool of development labor behind closed designs and proprietary rights claims. In most accounts of these events, as found in the trade press and on community web forums, the role of the villain is assigned to MakerBot Industries. The company's decision to close access to its products was widely deplored by members of the open hardware movement. Not only did it offend community norms about openness and sharing, but it also contradicted the pledges that the firm had made up to that point. However, recuperation may work in more insidious ways than by overtly hostile attempts at enclosure of the information commons. MakerBot Industries got away with its misconduct due to it having been in preparation for a much longer time, through a general displacement of community norms in the RepRap project. We have stressed the value of "repressive diversity" encoded into the Darwin side of the hyphenated "Darwinian-Marxism," with its *enrichissez-vous* undertones. It sapped the cohesiveness by means of which the community could otherwise have sanctioned violations of the GPL license and maintained a steady course toward crafting a self-reproducing, universal constructor.

It is apt to make a comparison with the previous case study of the Ronja project. In both case studies, a boom in market demand for the product in question forced the respective communities to make a collective decision about where to go next. The instigator of Ronja, "Clock" Kulhavy, stood firm with the original philosophy of a user-controlled technology. Consequently, he was isolated from other factions within the extended Ronja community who forked the project and started businesses around proprietary FSO technology. We grant that this is not a reassuring conclusion to draw. Still, the Ronja project withstood the test of recuperation, in the sense that the forks did not result in any marketable innovations or a valorization of capital. The Ronja technology remains publicly accessible, and although the development process has stalled, activists concerned about communication security and surveillance still draw inspiration from the example it set.

Another preliminary conclusion from the comparison is the contingency of leadership style in determining the direction of a hacker project. This contingency is, however, qualified by the importance of timing and location in deciding who will end up as instigator and community leader. By saying "timing" and "location," we allude to the framing conditions within which a single hacker project orientates itself. Ronja was launched at the moment when the free software movement stood at the zenith of its influence and prestige. Free software furnished the free FSO project not only with software tools and methodologies, but equally with values and a political analysis. RepRap was instigated some years later, at a time when hacker culture as a whole had drifted away from the ideologically stern program of freedom toward a more instrumentally oriented discourse about "openness." This observation corresponds with what we referred to in the theory chapter as the second time horizon over which recuperation processes unfold. Furthermore, it is at this analytical level, more so than at the first level of the individual project, that we can take the measure of the implications of the RepRap project being subsumed by capital and, along with it, the whole development stream of distributed, personalized manufacturing.

Concurrent with the commodification of desktop 3D printing is an emerging division of labor between different classes of users. In the theory chapter, we proposed such a taxonomy: communities of peer developers, crowds of users, and clouds of click workers. This point can be illustrated with reference to the different meanings given to the core idea of the RepRap project: to employ an existing fleet of 3D printers to make more such machines. At first, the concept of self-reproduction was organized on the basis of a moral economy within a community of hackers. The system of norms that supported this moral economy was undone by surging consumer demand and a corresponding speculative bubble in printed parts. Hence, eBay became the central distribution mechanism for these printed parts, concurrently providing a source of income for self-employed, relatively skilled, and independent users. Finally, the method of self-replication was rationalized by start-up 3D printing companies to overcome bottlenecks in production and reduce costs.

One example of this last step in the rationalization process is particularly interesting because it actualizes the labor question, namely Maker-Bot Industries' budding initiative to entice its former customers onto the company's assembly line. Effectively, the company experimented with an updated version of the old putting-out system. These customers were referred to in the firm's communiques as a "community of makers," but those "makers" had no say over the direction of the overall production process or the purposes to which it was put. Hence, we deem it more accurate to refer to them as a "cloud" of piece-rate workers. In contrast, the developers in the RepRap 3D printer community enjoyed autonomy over the assignment of work tasks and in determining the future direction of their collective existence. In the absence of such autonomy, the RepRap community could not have bootstrapped a thriving new market in consumer-grade desktop 3D printers, to which the start-up firms owed their existence. This is suggestive of the future of manufacturing, when communities of peer producers, crowds of users, and clouds of click workers will be put to work in a systematic fashion at different stages in the production process of physical goods.

5

HACKERSPACES
MEMORY AND FORGETTING THROUGH
GENERATIONS OF SHARED MACHINE SHOPS

In the previous chapters, we have discussed how informational capitalism responds to critique by implementing the demands of critics while simultaneously subverting the meaning of those demands. This is what we call recuperation. Even so, the process of recuperation is open to contestation and contingency. In the case of the Ronja project, analyzed in chapter 3, recuperation was successfully resisted, but at the price of the development process stalling when the community imploded under the weight of factional conflicts. In the case of RepRap, analyzed in chapter 4, the community's failure to resist recuperation is demonstrated in the emergence of a consumer market in desktop 3D printers. However, the organizational idea at the heart of the RepRap project, to distribute manufacturing capacity to society at large, proved more difficult to assimilate under a logic of commodity production. The decisive factor was the organizational difficulty of establishing managerial authority (also known as quality control) over a distributed production network. Hence, only the product innovation was successfully recuperated, while the potentially more disruptive idea of the RepRap project, to introduce a life-like logic of exponential growth and evolutionary change into the manufacturing process, proved unbending to the requirements of capital. This far into the argument, we need to substantiate the claim that recuperation processes form the overarching setting for any particular hacker project with its associated

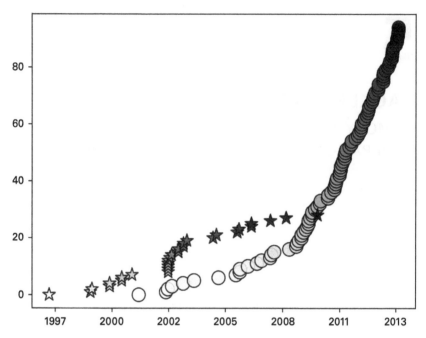

Hacklab vs. hackerspace domain name registrations over time (work of the authors). Stars mark new hacklab websites and circles mark new hackerspaces websites.

line of product developments. Recuperation acts concurrently on the legal, cultural, and technical landscapes that frame individual development projects, aligning those framing conditions with the requirements of state and industry.

This is precisely what we set out to demonstrate in this chapter, through a comparative, ethnographic study of the historical succession of different genres of "shared machine shops." Shared machine shops is not a practitioners' term. It was introduced by one of the authors of this book in a different context (Troxler and Maxigas 2014). Since then, it has gained some traction in the academic literature (Bosqué 2015; Dickel and Schrape 2017; Foster and Boeva 2018; Wenten 2019; etc.). Under this umbrella term, we gather a range of distinct but related phenomena that the practitioners variously refer to as makerspaces (Davies 2017; Lindtner 2015), FabLabs (Troxler 2015; Kohtala and Bosqué 2014; Gershenfeld 2005), Tech Shops (Hurst 2014; Schneider 1998), Men's Sheds (Wilson and Cordier 2013), incubators (Lindtner, Hertz, and Dourish 2014), hacklabs, hackerspaces, or accelerators (Cavalcanti 2014; Maxigas 2015). We will focus on the latter three subcategories of shared machine shops. They qualify for the denomination insofar as all three of them are physical locations where hackers, in addition to a more variegated assortment of practitioners, gather to socialize and collaborate on technology projects. Crucially, participants in hacklabs, hackerspaces, and accelerators all draw on hacker culture in order to distinguish themselves from other kinds of entities and subcultures. Furthermore, as will be clear from the discussion below, they have a historical lineage in common.

We home in on these three genres of shared machine shops due to their suitability for clarifying some analytical points about recuperation. Hacklabs showcase the most politicized space, as well as being the first instantiation of a shared machine shop. The other genres were spawned from this point of origin. Next in turn are hackerspaces. The activist credo of hacklabs has here been supplanted by an almost exclusive focus on tinkering with technology, concurrent with the promotion of the ethos of sharing and repairing. Jumping over a couple of intermediary steps, we close the discussion with a recent manifestation of shared machine shops, called "accelerators." These are imbued with a rhetoric of job creation, urban and regional development, entrepreneurial exploits, and high-tech innovation. The three genres represent successive historical steps in a progressive logic of recuperation.

In the analytical scheme introduced in the theory chapter, the movement of shared machine shops is located within the second time horizon, situated in between the life cycle of an individual hacker project (first time horizon) and the epochal transitions in capitalism as an evolving whole (third time horizon). The intermediate level of shared machine shops is due to the fact that they provide a material infrastructure and cultural context within which individual hacker projects may run their course. It bears stressing that shared machine shops are key sites for the reproduction of hacker culture, both in the cities where they are located and at the aggregate level. Shared machine shops, alongside annual conferences and outdoor camps, are the physical places in which members regularly gather and confirm their own and others' belonging to the same movement.

Through such ritual gatherings, a compact social formation emerges that fills the otherwise hollow and abstract notion of "community" (Coleman 2010). In the process, common meanings are established and attention is drawn to matters of concern, potentially triggering collective action. What kind of identity a space adopts is therefore decisive for how much room for maneuver will exist for critical engineering practices—or, alternatively, how expediently such practices will be turned into technological and organizational innovations for capital.

By adopting a longer historical perspective on shared machine shops, our investigation extends across generational shifts within hacker culture. Thus, we want to stress the importance of shared historical memory. At stake is the transmission of the lessons learned from past events and engagements. What aspects of the past are successfully passed on from one generation to the next, and what aspects are omitted, is subject to incessant internal contestations and pressure from external actors. Later genres of shared machine shops feed off the imaginaries, legitimacy, and technical achievements of their forerunners. Concurrently, the different genres have come to enact very different political-economic agendas, something that we will demonstrate in relation to urban regeneration plans. In this, we find a telltale sign of recuperation.

Traces of an ongoing recuperation process can also be detected at an aesthetic level, such as, for instance, in the prevalence of graffiti on the walls of a physical space. This connects to the remark we made in the introduction with a reference to Immanuel Kant's philosophy. It takes an aesthetical judgment to spot recuperation attempts. To substantiate this claim, we will conclude each section with a description of the patterns of indoor lighting that are typical in different genres of shared machine shops. Here we take a cue from satellite images of the night-time earth. In popular culture, the uneven distribution of light emissions displayed on those photos is taken to represent centers of economic activity versus areas of nonactivity. In the following pages, we venture to apply the same reasoning to shared machine shops. The intensity and rhythm of the light emitted from a physical space indicates the extent to which the space in question has been assimilated into the regular office hours of the surrounding society.

THE PREHISTORY OF SHARED MACHINE SHOPS

The proposal that we made at the outset of this book, with reference to William Morris's allegorical tale of a peasant uprising, namely, that hackers are caught up in a whirlwind of social forces that can be meaningfully linked to a longer history of industrial conflicts, finds empirical support in the precursors to shared machine shops. The onset of deindustrialization in Western countries during the 1970s created the material conditions for a wave of social mobilization around the idea of a democratization of manufacturing capacities. A palpable, material trace thereof are the machine tools that many hackerspaces have inherited from defunct manufacturing firms in their vicinity. In many cases, the physical space itself is housed in the remnants of a former manufacturing building. More to the point, however, is that the very notion that the means of production can be reclaimed by setting up a shared machine shop "outside of the factory gates" is an outcome of defeats in industrial conflicts of the past.

Numerous authors before us have connected the movement of shared machine shops to the Lucas Plan in England and the subsequent promotion of technology networks for socially useful production by the Greater London Council (Smith 2014; Smith et al. 2017). Similar developments took place simultaneously in Scandinavian countries under the names Collective Resource Approach and participatory design (Ehn, Nilsson, and Topgaard 2014). Drawing support from trade unions, socialist parties, and universities, these interventions into the design process sought to strengthen organized labor in anticipation of the automation of industry. These initiatives have been documented and analyzed elsewhere, so we will give only a brief account of them here. The discussion serves to demonstrate the analytical benefits of studying hackers within the second and third time horizons—that is to say, to diagnose the trends of developments within a larger movement of hacker projects and, furthermore, to relate this trajectory to capitalism as an evolving whole. An unexpected outcome of deindustrialization and automation was that it set the scene for hacker culture to flourish.

Our interpretative framework resonates with Harry Cleaver's concept of "cycles of struggle" (2017, 58), as well as its application in the context of information capitalism by Nick Dyer-Witheford (2015). The concept describes how forms of social contestation coevolve in tandem with the ever-shifting regime of capital accumulation. Technical innovations are interpreted as capital's response to working-class resistance. A major restructuring of industry during the 1970s, coupled with automation, outsourcing, and financialization, allowed capital to dissolve the strongholds that the Fordist mass worker had mounted inside the production apparatus. The technical and organizational composition of the working class was thus also dissolved, along with its former identity constructions. A token of this is the difficulty of recognizing the present-day experiences of the working class as belonging within that old Fordist register.

On the basis of this conceptual argument, we advance the proposition that the rise of shared machine shops is not merely an unexpected, secondary outcome of the defeat of the Fordist mass worker. It concurrently inaugurates a new cycle of struggle, corresponding to the new cycle of capital accumulation. The struggle of hackers showcases how the extraction of value, as well as contestations against this extraction process, takes place outside of the contractual employment relation and the legally recognized "abode of production."

In 1976, during a period of crisis for the whole of the British manufacturing industry, the military contractor Lucas Aerospace faced the prospect of insolvency. Workers took the initiative and drafted an alternative business plan to save their jobs while steering the arms manufacturing capacities of the firm toward "socially useful production." For posterity, this initiative has become known as the Lucas Plan. The proposal included market research and blueprints for more than a hundred socially useful products. Coupled with the alternative design items was a reconceptualization of the labor process to involve the workers in strategic decision making. Representatives of the Lucas Aerospace Shop Stewards' Combine Committee toured the country with one of their prototypes, a bus-train hybrid vehicle, to promote the idea to the public. Despite these efforts, the Lucas Plan stalled due to fierce resistance from management, the government, and the upper ranks of the trade unions. After its demise, social movements drew inspiration from the initiative and began to rally around the idea of involving a broader range of actors in the decision-making process over technology development. In a retrospective analysis of the Lucas Plan, Adrian Smith concludes that it "challenged fundamental assumptions about how design and innovation should operate" (2014, 2).

Next, the demand for socially useful production was championed by a motley coalition of left-wing Labour politicians, neighborhood community groups, peace activists, and the nascent environmental movement. The authoritative account of these movements was given in the book *Architect or Bee? The Human/Technology Relationship* by Mike Cooley, the key promoter behind the Greater London Enterprise Board (1987). At the moment when Margaret Thatcher was ascending to power, the Greater London Council asserted itself as a holdout of labor power. The economic policies of the national government were resisted at the municipal level. It was within this highly conflictual political climate that the network of community machine workshops was inaugurated on London's outskirts. The invitation to local residents to become involved in the creation of alternative technology was framed as a challenge both to managerial authority on the shop floor and neoliberal government policies. The stress placed on "alternative" pathways in technological development was taking aim at the rhetoric of technological and economic determinism, succinctly put in Margaret Tatcher's slogan: "there is no alternative," which underpinned the ideological scaffolding of neoliberalism in those early days (Greater London Enterprise Board 1984, 34).

Examples of items produced in the neighborhood workshops include "small-scale wind turbines, energy conservation services, disability devices, products made from recycled materials, toys for children, community computer networks, and a women's IT co-operative" (Smith 2014, 1). These were documented and registered in an open-source database as the potential basis for co-operatives, founded with support from the Greater London Enterprise Board, while the movement spread to other cities. The workshops became a hub of activity linking a variety of actors who were invested in the idea of socially useful production. The eventual decline and demise of the movement coincided with the consolidation of Thatcherism, epitomized by the suppression of the miners' strike in 1985. The Greater London Council, led by Ken Livingstone who championed the movement, was abolished the following year.

This all suggests that the birth of the idea of "socially useful production" and neighborhood-controlled machine workshops were closely linked to the fate of the British working class at the onset of deindustrialization. Proposals of the same kind surfaced in other countries where the working class faced similarly bleak prospects. Academics working in liaison with trade unions launched the participatory design movement in Scandinavia and the science shops movement in the Netherlands. These initiatives were prompted by the anticipated social consequences of a nascent technology that, back then, was mostly associated with factory automation, namely computerization. The participatory design movement aspired to involve workers in the decision-making process over information technology and the design of human–computer interactions. A case in point was the digitization of typesetting (e.g., "the transition from lead composing to computer-based text processing and phototypesetting"), where the participatory design approach figured as an "intervention into managerial and technical plans for manning, education, investments, etc.," rooted in "Marxist labor process theories and the practical experience of the investigation groups" in which workers participated (Ehn 1988, 12). Retroactive accounts, such as that by Kraft and Bansler (1994), suggest that, in spite of the good intentions of the academic researchers, these interventions achieved limited success.

The underlying concepts, along with the associated political interventions, underwent the familiar dialectic of critique and recuperation. Here, we can only briefly trace this trajectory. Some inflections in the terminology are, however, suggestive. "Cooperative design" turned into "participative design," later to be reinvented as "co-design." The centrality of workers in the design methodology was gradually pushed aside and replaced with a more varied assortment of "users" and "stakeholders." Whereas the original proposals of participative design had been anchored in a society-wide analysis of the class antagonism between workers and management, later reformulations sought to foster consensual work relations by identifying cost-efficient solutions to remove direct sources of employee discontent.

The science shops movement followed a similar trajectory. In the early days, the case for democratizing science was made as an intervention on the side of labor in the class conflict. Scientific knowledge production was understood to belong to the arsenal of managers, while trade unions had far fewer resources to pool into methodological research about hazardous working conditions, the benchmarking of productivity, strategic foresight, and so on. Science shops sought to remedy this imbalance by elevating trade unions to clients of research output. The topics of PhD theses, dissertations, and research projects would be formulated and implemented in collaboration with trade unions and factory workers.

The first science shop was established in 1977 at the University of Amsterdam on the initiative of the Department of Science Dynamics. Another notable project included the consultation on an electronic payment system involving the national post office, clerical workers, and service companies (Leydesdorff and Van Den Besselaar 1987). A retrospective evaluation by former participants concludes that "the disappointments over the role of experiments" were due to the declining power of the unions, which could not muster organizational or research power to match that of the parties on the other side of the negotiating table (Leydesdorff and Van Den Besselaar 1987, 156). The critique of scientific knowledge production in the service of capital was duly recuperated at the moment when broadened participation in decision-making processes became a way of justifying existing work relations. A case in point is the dialogue panels on science and consensus conferences, staged by the European Commission to bestow legitimacy on its business-friendly, laissez-faire innovation policies (Horst and Irwin, 2010; Kelty 2019, 156–157).

The cases mentioned above, the Lucas Plan, the municipal network of neighborhood community workshops, the participatory design initiative, and the science shops movement are precursors to the subsequent wave of shared machine shops. Although the older examples are, biographically speaking, disconnected from hacker culture, they converge thematically in the vision of democratizing scientific knowledge production and technology development. This vision was first articulated in the context of industrial conflicts during a period of crisis and structural reform. The all-important difference between these two periods is that the initiatives of the first period sought to challenge capital's prerogative over science and technology from within established institutional arrangements. Although the experiments in bottom-up decision making often clashed with union and party leadership, whose resistance proved decisive in many instances, the whole undertaking was nevertheless framed by trade unionism and electoral politics. The decline of both of these institutions in the following years concurrently spelled an end to the experimentation taking place at their fringes. In Cleaver's assessment, for cycles of struggle to be effective, they must eventually break loose from and prevail independently of the circuits of capital: "The primary implication that I draw from what I have sketched is how revolutionary struggle must involve not only the rupture of the circuits of capital but escape from them through both their destruction and the creation of alternative social relationships. Doing so certainly requires recognizing the patterns of those circuits as well as their content—the endless imposition of work and the subordination of life to work. But such recognition is only meaningful if it informs rupture and creation" (Cleaver 2016, 23).

When hackers challenge capital's prerogative over science and technology, the challenger is located outside of contractual wage relations and occupational identities. The outsider position of the hacker vis-à-vis formal institutions modifies the terms of conflict. Inside the confines of the employment contract, managers and employees are locked into antagonistic opposition over what expenditure of time and effort is deemed equivalent to a fixed amount of remuneration (i.e., wage). This is not an issue for hackers, who dispense their creativity freely on a nonremunerative basis. Instead, strife in the computer underground circles around keeping the results of their collective endeavor within the information commons. The information commons has to be protected because it allows hackers to hack, serving as a resource pool of ideas, standards, code, data and tools. In other words, it is part and parcel of the material conditions that allow them to reproduce the social relations that make them hackers. This interpretation of hacker culture resonates with the stress put on worker autonomy by Cleaver and others within the same branch of Marxism. Hacker communities have established themselves as autonomous sites of value production. The reverse side of the same trend is capital's reorientation toward capturing value across a heterogeneous field of activities and interactions. Hence, hackers confront the threat of seeing their collective existence as hackers subsumed under capital and optimized to the needs of an open innovation model.

As will transpire from the discussion below, the outsider position of hackers in relation to both capital (contractual employment) and the political institutions of the Fordist mass worker (trade unions, socialist parties) gives no guarantee against them becoming incorporated into institutional arrangements of a different sort. With each successive genre of shared machine shops, both the concept and the practices have become increasingly aligned with the requirements of state and industry. Further down in the text, we argue that this trend can be detected from the different roles that different genres of shared machine shops have played in urban regeneration plans and real estate development. When the history of shared machine shops is recounted within the second time horizon, it is a story about decline. In the frame of the third time horizon, however, the story is about one cycle of struggle being supplanted by another, corresponding to a period of restructuring of capitalist processes of accumulation.

HACKLABS

With the term "hacklabs," we are referring to the first generation of shared machine shops that drew inspiration from and linked up with hacker culture. Many hacklabs started out as media technology hubs inside squatted social centers. Geographically, they were concentrated in Italy and Spain, where the squatting scene was particularly strong. Hacklabs provided technical support for demonstrations and day-to-day activities at the social centers. A leading theme of the setting was the idea of "territory of autonomia" (Wright 2002). In the native language of these movements, the link to workplace struggles was explicit, although this is not explicit in the English translation. What are referred to as "occupied social centers" in English go by the name of *Centro Sociale Okupado y Autogestinado* in Spanish or *Centro Sociale Occupato Autogestito* in Italian. The last word stands for "self-management."

A milestone in the organization of hacklabs as a social movement in its own right was the 1999 hackmeeting at the Centro Sociale Occupato Autogestito in Milan, known as Deposito Bulk (ana 2004; Anarchopedia contributors 2006; "storia" 2010). During a three-day session, the social center brought together politicized hackers, media activists, and squatters, who put on a program that mixed a festive spirit with hands-on experimentation, as well as workshops, talks, and debates. The gathering was underwritten by the pirate-anarchic ethos of Hakim Bey's *Temporary Autonomous Zones* (1991). The notion strongly resonated with practices and imaginaries of independent media activism located in occupied social centers. The "T.A.Z." concept, as it was often stylized, celebrated the precarious conditions of these settings. Freedom nested for a brief moment in the inevitable cracks and crannies within large, oppressive, and overly complex systems of domination. Adding to this ideological inventory were celebrations of nomadic life and rhizomatic networks (i.e., the insignias of Deleuzian philosophy).

Participants at the hackmeeting in Milan, dirty and tired but thrilled to be part of a Temporary Autonomous Zone, decided to render the experience permanent by establishing hacklabs in occupied social centers in their home towns. Isolated experiments that had already existed, such as Freak-Net in Catania, Sicily, provided inspiration and stability to the emerging network of hacklabs. As Bazzichelli observed some years later, "cyberpunk in Italy has taken on the connotations of a political movement" (2009, 68). Annual hackmeetings and local hacklabs constituted the spatial-temporal dimensions of political cyberpunk (Bazichelli 2009, 68–75).[1] The network of hacklabs gradually spread from Italy to other countries with a strong tradition of squatting and autonomous politics, although its stronghold remained in southern Europe, where Italian and Iberian hackmeetings continue to this day.

Hacklabs ran on recycled hardware, bootlegged internet access, and free software. In the spirit of the social centers, which catered to their neighborhood communities with cultural happenings, vegan kitchens, daycare services, and so on, the hacklabs provided free public access to computers and the internet. In addition to providing computer access for the public, hacklabs typically organized workshops, ranging from teaching basic computer use and staging GNU/Linux installation parties to educating social movement activists in computer security and cryptography (Yuill 2008, para. 7). In return, the squatters offered the basic material conditions for the hacklab to exist in the first place, a residence, often located in a prime metropolitan area where rents were extortionate. More than just letting a floor in an empty building, the squatters had to protect their hacklab from police raids and incursions by fascist groups.

Thanks to their embeddedness within a social movement with an anarchist and autonomist outlook, the hackers were furnished with ideological and political training. There was a natural connection with the underground media activities taking place at most social centers; in particular, the broadcasting of pirate radio. Hackers and media activists converged in their defiance of intellectual property, and file sharing presented itself as the latest addition to the pirate arsenal. Like the pirates of legend, squatters legitimized trespassing and appropriation through references to artistic freedom and social redistribution. The aesthetic of the remix, repurposing, and bric-a-brac lent itself easily to trafficking in intellectual property and the reclamation of real estate.

The aforementioned combination of informal sociality, political activism, and hands-on engineering developed into a methodology. A late example is the Hackafou hacklab, which hosted the 2012 Iberian Hackmeeting. This hacklab was situated in Calafou, an old industrial village near Barcelona, turned into a "Post-capitalist, Eco-industrial Colony" by former squatters, where one of the authors lived and participated in the 2010s. In the heyday of this hacklab, hackers and activists could be found there at any hour of the day, coding stoically on flotsam laptops and Frankenstein desktops, falling asleep, exhausted, across a table or in an armchair, throwing a party or playing video games. Hacking was a way of life for many of the participants.

Hacklabs became integrated into the wider hacker culture through a radical reading of free software ideology and practices. The inventor, practitioner, and evangelist of free software, Richard Stallman, articulated a powerful and detailed practical and theoretical critique of intellectual property, arguing that licenses should protect the rights of users rather than producers, while insisting that developers could still profit from software production in various ways. His views have been interpreted as advocating communism—even if sometimes ironically—by a wide range of commentators, from Microsoft CEO Bill Gates (Stallman 2015) to labor activist and political theorist Richard Barbrook (2015). Over the last few decades, Stallman has been touring the world on a never-ending crusade against proprietary software, including appearances at hacklabs and hackmeetings in Spain and Italy.[2] Free software was a key component in the ideology and practices of hacklabs. This led to cooperation and conflict with both the free software movement and the squatter movement. Hacklab participants saw a direct connection between the critique of intellectual property articulated by Richard Stallman and the critique of real estate speculation that legitimized occupied social centers. Events in squats, such as hackmeetings, were supported by donations, in the spirit of another staple of occupied social centers, freeshops. In the freeshop, visitors were offered clothes and other items free of charge.[3]

The alliance between hacklabs, squats, and free software was tested during an event in 2005 that brought a prominent cross-section of free software developers and anarchist hackers into direct contact with each other. *Rencontres Mondiales de Logiciel Libre* (RMLL), alias Libre Software Meeting, is an annual meeting of free software developers, enthusiasts and users, which combines a conference for the community with outreach promoting free software to a larger audience. When it took place in Dijon, its evening program was hosted by the PRINT hacklab in the Les Tanneries occupied social center under the title *Nocturnes*. This was a major opportunity to present the anarchist reading of free software to its core participants, advocating an anticapitalist critique of private property as a totality, instead of intellectual property as a specific aberration of the capitalist mode of production. According to recollections in interviews and published reports, the event took both sides by surprise, leaving many free software developers perplexed and some scared, while the hosts were disillusioned by the bourgeois and conservative attitudes of their guests.[4] While hacklab participants attempted to radicalize free software developers, their main efforts were directed toward evangelizing free software amongst squatters and anarchists, through setting up network connections and public terminals in squats and providing tech support to the local community. Activists could better identify with the radical reading of free software licenses and appreciated the practical benefits of free software in terms of cost, flexibility, and security. However, the relationship was not without tensions. A string of actions highlighted the frustration of hacklab participants with the slow adoption of free software operating systems within the local social centers.

The Escamot Espiral (the Swirl Commando) was set up by hackers from the Kernel Panic Hacklab in Barcelona. On February 22, 2007, they marched in a demonstration from their base in the La Quimera occupied social center, wearing black balaclavas bearing the eponymous red swirl logo of Debian, the leading community-oriented GNU/Linux operating system distribution, complemented with cheeky devil horns. They stormed into La Torna in the Gràcia neighborhood, an allied Catalan independentist community center. The Commando interrupted the meeting and read their manifesto, declaring that "we're fed up of being 'those Linux freaks,'" whose arguments about the social justice and coherence inherent in free software are not heard by the wider movements. They told the assembly of independentists that "enough is enough of alibis. There is no more effort in switching to GNU/Linux than facing the police in demonstrations, [and] enduring evictions." Therefore, they announced to the terrified audience that "we're stepping into direct action: any Windows computer we find will be immediately and mercilessly converted to GNU/Linux," and proceeded to install Debian on the public terminals in the place, wiping out the previous Windows operating system. The action was a light-hearted but strongly felt reminder to their comrades that the cause of free software should be taken as seriously as the other struggles that participants stood for, from free public spaces through to vegan food and Catalan independence. This and two subsequent actions[5] demonstrated that hacklabs supported social movements in free spaces, but also brought their own political analysis to the mix.[6]

Beyond the social and political context, the particular configuration of hacklabs was well adapted to the specific media landscape at the time. In addition to the free software that was ubiquitously used in the hacklabs, IBM-compatible PCs, modems, and wireless routers were part of the standard inventory. The electronics were salvaged from trash bins and occupied buildings and used to build up self-managed infrastructures. The modularity of the IBM-compatible desktop computers lent itself to engineering practices articulated in a context where recycling and stealing were the primary means of accessing resources. Of particular note are wireless routers, which had started to circulate on the retail market as a means of extending connectivity at corporate trade fairs and in office buildings. Yet, 2.4-GHz waves seeped through walls and onto the streets, so that wireless routers could be repurposed to build public networks available to everyone. Community wireless networks, discussed in the chapter on Ronja, spawned in the social centers, and moved outward from there, often making up the backbone of a city's wireless communications network. Similarly, civilian-grade WEP encryption was used to interfere with the boundaries of the bourgeois private sphere. The same connectivity could provide an uplink for public networks through widely available exploitation of their flawed algorithm. Such exploits were celebrated as a material demonstration of the feasibility of the glitch strategy offered by political cyberpunk. More than anything else, it was the ubiquitous network cables that characterized the visual outlook of these scenes, often doubling as ropes in electrical installations and building material for barricades.

It proved much harder for the hacklabs to accommodate themselves to the next layer of infrastructure that was rolled out on top of their native environment—in particular, smartphones and social media platforms. With the consolidation of corporate power over communications technology, the cutting edge of mainstream media consumption moved away from those practices in which hackers excel, recycling and repurposing. Ultimately, however, it was political decisions and not technological changes that forced the hacklabs into retreat. The decline set in when the squatting movement came under intensified state repression. New eviction laws and toughened police tactics put the squatters, and with them, the social and material basis of the hacklabs, under severe stress.

At a time when the autonomous social movements were suffering defeats and their ideological underpinnings were waning, political cyberpunk paradoxically came to the rescue, lending them another narrative of justification, if not hope. It is paradoxical because it took a dialectic reversal to draw inspiration from a literary genre that distinguished itself by its dystopian approach to the future, characterized by cynicism and pessimism. The combination of Gothic sensibilities, hard-boiled film noir, and science fiction (Whatley 2013; Rapatzikou 2004) proved ideal for a popular movement in retreat. According to this narrative, the survival—if not success—of self-organized collectives depended on the unintended consequences, systemic errors, and chaotic entropy inherent in capitalist progress. The affective, visual, and aural substance of such a configuration was expressed through the aesthetic of the glitch, for which the found materials that filled hacklabs and the broken infrastructure that surrounded them were ideal material.[7]

Hacklabs were often dimly lit, for both objective and subjective reasons. Objectively, the light installation was one of the many things participants had to take care of on their own. Everything from sourcing electricity, through scavenging cables, to fashioning lampshades were in the hands of the occupants. A staple of squat illumination was Christmas tree lights, especially on staircases, where they could be hung to illuminate several floors from a single socket. The importance of this consists in the fact that sockets were often a scarce resource in the ramshackle buildings. While one of the sites for the ASCII squatted internet cafe in Amsterdam was in a prime corner location, with a shopfront façade on the ground floor, after the next iteration of the eviction-and-occupation cycle, the same collective ended up in a basement. Subjectively, many preferred the darker environments depicted in hacker and cyberpunk movies. Flickering light bulbs due to jittery electrical connections expressed the transitional and liminal nature of temporary autonomous zones. In essence, the limited illumination of hacklabs was not accidental: it stemmed from the material and psychological conditions of the occupants.

HACKERSPACES

The birth of hackerspaces as a social movement can be dated to 2007, when Jens Ohlig and Lars Weiler presented the "hackerspace design patterns" to visitors from the United States, whom they were hosting at C4-Labor, the Chaos Computer Club Cologne Laboratory. The Chaos Computer Club was founded in Hamburg in 1981 as "a galactic community of life forms, independent of age, sex, race or societal orientation, which strives across borders for freedom of information." Over the years, it grew into the largest hacker organization on the planet, with over 7,700 members and many local chapters as of 2021. Many German hackerspaces emerged from the office and club spaces of these chapters. As recollected in a dedicated book published soon after the legendary events of 2007, "In 2007, a number of meek and lonely hackers from the States went on the Hackers On A Plane adventure going to Chaos Communication Camp and then travelling around Europe visiting hackerspaces. When they arrived at C4 in Cologne, Jens Ohlig and Lars Weiler gave the first presentation of the Hackerspace Design patterns. It's a document made with the wisdom of doing it wrong in so many wonderful and disastrous ways" (Bre and Astera 2008, 92).

The two went on to speak about the recipe for hackerspaces at the Chaos Communication Camp that summer[8] and later at the prestigious annual Chaos Communication Congress at the end of the year (Ohlig and Weiler 2007). However, the birth of a new movement had already been heralded in the canonical medium of US mainstream cyberculture, *Wired Magazine* (2007), with John Borland reporting from the hacker camp in August. Pioneering spaces such as the c-base in Berlin (established 1995), the aforementioned C4-Labor (established 1998), and Metalab in Vienna (established 2006) served as inspiration for a consistent genre to emerge.

Even before the establishment of the first hacklabs and hackerspaces, a whole range of important and inspirational physical spaces dedicated to the cultivation of cyberculture existed in Europe. A good example is Public Netbase (1994–2006) in Vienna, which was closed by the Haider government in the year of Metalab's founding, or the Mama Multimedia Institute in Zagreb, Croatia (2000–), which put artists, activists, and geeks in contact with each other and still hosts a hacklab. However, these isolated sites never came together into a popular genre as an ideal-typical social formation that exhibits a high level of consistency across many instances. It is the latter that we are investigating under the rubric of shared machine shops. Such a common understanding was reached at the annual hacker gatherings. It resulted in the proliferation of hackerspaces in northern Europe and North America. The pattern spread to other parts of the world in subsequent years, leading to more than a thousand active spaces today (Hackerspaces Wiki contributors 2020b; Murillo 2019; Davies 2017).

The Chaos Computer Club was established as an activist organization (Denker 2014). It continues to play a role in policy making and enacts high-profile techno-political interventions (Kubitschko 2015). However, the self-definition of hackerspaces at the hackerspaces.org aggregation website highlights tinkering with technology as the sole purpose of hackerspaces: "Hackerspaces are community-operated physical places, where people share their interest in tinkering with technology, meet and work on their projects, and learn from each other" (Hackerspaces Wiki contributors 2020a).

Such a definition is good for identifying a common thread that unites hackerspaces and stands up to empirical scrutiny based on our fieldwork experiences. Whereas many other types of shared machine shops use similar phrases to define themselves, their visitors often find little more than a few machines that identify the innovation potential of the space, coupled with participants writing grant applications on MacBooks. In contrast, hackerspaces are filled with evidence of hardware hacking and other technological experimentation. Members are also more inclined to engage with the materiality of media. However, this definition papers over the diversity of participants—a key aspect of their attraction as sites for community building around technological issues.

That keyholding members will have 24/7 access is taken for granted in the hackerspace genre, but there is also a complementary sense of being a public space in a remarkably broad definition of the concept. For instance, at the aforementioned c-base in Berlin, which is fashioned to resemble a grounded spaceship and alien archaeological site (Fichtner 2015), visitors are DNA scanned at the entrance gate by a mock-sci-fi machine, while a myriad of diverse silhouettes flash up on the screen, only to be told "greetings, human" in a synthetic voice. A corresponding wiki page states that "the lock and the main hall are available to all forms of life" (c-base wiki contributors 2019, para. 27).[9] These gatekeeping practices—when interpreted in the wider context of hacker mythology and c-base architecture—manifest antiracist and antispeciesist values, veiled in an ironic commentary on surveillance and control, modernism and progress. It is no surprise, then, that in strict sociological terms, such cosmopolitan—or more accurately, "universal"—aspirations fall short of the reality, as we discuss in more detail later. This style of inclusion draws a largely white, male, able-bodied, middle-aged, city-dwelling, and well-educated crowd.

Meanwhile, "door systems" at many European hackerspaces provide a mechanical and symbolic measure of participation and productivity.[10] At Technologia Incognita in Amsterdam, upon entering the space, one faces a big red button at the center of a contraption. Pushing the button lights up eight LEDs around a ship's wheel drawn in silvery solder on a circuit board the size of a business card. The same circuit board doubles as an educational device that can be purchased as a kit from a vending machine and soldered together according to instructions.[11] On the door system installation, the push button is further connected to a single board computer that announces on the internet that the space is open. This is achieved through an API (application programming interface) protocol standardized and aggregated across hundreds of hackerspaces. The real-time opening data is used by various other machines to inform users and gather statistics.

For example, the front page of the hackerspace website now spells "TechInc is OPEN." Statistics on past opening and closing times are updated on another public web page, and a chatbot announces the event on the IRC channel. Once a key-holding member pushes the button upon entering the hackerspace, unannounced visitors are welcomed to come or go as in any other public space. The door system resembles a time clock at the factory gates, but this gatekeeping device is made to serve a very different purpose. The factory clock enforced the contractually stipulated length of the working day upon employees. The open-door system is designed to enhance the flexibility of the opening hours, so that members can come and go as they please, while allowing the hackerspace to stay open to the public as much as possible. The open-door system integrates hackerspaces into a single material infrastructure that performs the ideal of openness in hacker culture. Even though hackerspaces are organized along the lines of membership clubs, they often double as inclusive and accessible public spaces in their neighborhoods.

The open-door policy notwithstanding, there are numerous limits to the enactment of openness as a political ideal and a social norm in hackerspaces. Three limitations of the universalistic ideals of hackers can be readily detected: firstly, the marked lack of diversity in terms of professional trades and political backgrounds among the members; secondly, class divisions in the surrounding society resurface inside the hackerspace; and thirdly, the extreme gender imbalance, in comparison to which the prominence of queer identities is noteworthy. Although the topics of class, race, and gender have been much debated in hackerspaces and at hacker conferences, the overwhelming majority of participants are still male, white, and middle class (Davies 2017).

Firstly, the point about the diversity of professional trades and political backgrounds can be illustrated with a pair of ethnographic vignettes, originating in 2011 at the London Hackspace and in 2010 at H.A.C.K., the hackerspace in Budapest. One of the authors of this book paid regular visits to the London hackerspace while living and working in the city for an extended year, concurrent with the peak of the Occupy movement. One visit became emblematic of the range of characters who frequented the space. Two Occupy London activists worked on the digital infrastructure of the movement, fixing a wiki installation while discussing how it was being used to organize the camp at Trafalgar Square. The work session took place in the so-called "dirty room," the dedicated wood and metal workshop.

This opens onto the lounge, featuring an assortment of sofas in various conditions of (dis)repair and serving as the main hall of the hackerspace. The lounge hosted the only publicly advertised event that day: a "mind-hacking workshop" where two members were exercising hypnosis in order to broaden their imaginations and experiment with altered body states. The computer room next door was the scene of a private conversation between radio amateurs, one of whom gave a detailed account of his latest professional engagement—sourcing radio equipment for the Metropolitan Police and training the force in its use. In line with the definition above, all these people could be identified as technological tinkerers of some sort. Yet, their professional backgrounds and political attitudes were diverse and sometimes even contradictory. They were no doubt aware of the diversity and contradictions within the membership, since the radio amateur also discussed the presence of Occupy activists in the space. Interestingly, his immediate concerns were starkly practical rather than directly political. He defended the right of activists to use the workshop tools for building contraptions for the occupation but commented on the mess they left behind in the workshop and suspected that some hung out in the space as an alternative to sleeping rough. We return to this observation in a subsequent paragraph on class conflicts.[12]

The other memorable field experience occurred during a regular information security-themed meetup in Budapest. One of the authors of this book was present at this meeting by virtue of being a regular member and cofounder of the hackerspace. The informal gathering culminated in a heated debate about vulnerable software products used to prop up the digital security infrastructures of some more obscure part of the Hungarian state apparatus. The pizza and beer were sponsored by a small security company providing penetration testing and security certification services to major commercial players in the banking sector. The participants hinted at their backgrounds during the conversation, which went on late into the night. It sounded as though the people at the table included academic staff, secret service operatives, commercial technology consultants, and perhaps even Anonymous activists. What brought these otherwise warring parties together was their interest in information security, which made ideological differences irrelevant in that moment. This political agnosticism of hacker politics has also been observed elsewhere (Coleman 2004).

In fact, the class conflict in the hackerspace movement came to a head simultaneously with the Occupy and Anonymous movements, that is, in 2011 and 2012. Murillo describes the long-term community dynamics, friction, and turbulence that resulted in activists who were sleeping in the space having to call up "middle-class" members to help them out (2019, 213–214). These incidents suggest that Occupy activists and their allies made extensive use of free spaces, including available hackerspaces, which put additional strain on these communities and reinforced existing fault lines. Interestingly, some hackerspaces integrated homeless people into their communities for several years, even though it seems that such initiatives also eventually failed, given the institutional constraints on hackerspaces. "Chess players" have been recognized by hackerspace members from three otherwise very different hackerspaces (Noisebridge in San Francisco, Metalab in Vienna, and Mama in Zagreb, during interviews with Budington, December 26, 2016, Wolf, August 30, 2001, and Mars, April 12, 2014, respectively). This suggests that gatekeeping practices at hackerspaces support subjects who display hacker traits identified by Coleman (2014)—technical virtuosity and the performance of craftiness—regardless of class distinctions. However, Noisebridge later staged a reboot involving closing the space and changing the keys, while the Metalab assembly approved of a plan to convert the only shower into a dark room for developing analogue film, and changes in Mama meant that chess players went elsewhere. The consistency between three otherwise quite different sites—San Francisco, Vienna, and Zagreb—lends further support to the proposition that hackerspaces constitute a genre of their own: a single social formation that can accommodate a diversity of subjects, with notable limitations.

Thirdly, the lack of gender balance and the prominence of queer identities is an enduring feature of hackerspaces. Hacker culture is far removed from the macho mainstream of pop culture—instead it performs an alternative, geek masculinity. In line with the conflict resolution practices in open-source culture and free software projects, the movement forked through the establishment of nonmale—female-identified, trans-forward, and LGBT-friendly—hackerspaces to provide "safer spaces" for technological experimentation and mutual aid (Toupin 2013, 1; 2014). While women consistently report sexist behavior in hackerspaces, hacker scenes have been able, in contrast with other patriarchal milieus, to accommodate nonbinary identities. As hacker anthropologist Gabriella Coleman asks, "Why, for instance, are gender benders, queer hackers, and female trolls common and openly accepted categories, but female participation in technical circles remains low?" (2014, 175). We can answer this question based on our own ethnographic observations. The crux of the matter is that geek masculinity appears to manifest an effective critique of traditional male macho stereotypes, and perhaps even gender binaries as such, but falls short of questioning and subverting the hegemonic gender relations at the heart of patriarchy. This is why it may accommodate queer identities that appear to be compatible with its sensibilities, while failing to include cis-women: a serious political limitation.

Forays by hackerspace members and audiences into institutional politics include two prominent examples. One was the foundation and meetings of the German Pirate Party at the c-base hackerspace in Berlin. The party went on to electoral success, including sending several representatives to the European Parliament. The other was an earlier campaign against software patents coordinated by activists at a European level. It prevented the introduction of patents on algorithms, which had been successfully introduced in the United States. One of the authors of this book has previously argued that hackerspaces today constitute a particularly valuable, materialist approach to the wider project of democratizing science and technology—in sharp contrast, for instance, to consensus conferences (Maxigas 2013). The other author has argued in a different context that, under certain circumstances, hackers' faith in technological determinism can boost their morale and fortitude, especially when facing up against a stronger opponent (Söderberg 2013). Taken together, these arguments indicate that the political disengagement of hackerspaces is ambiguous. Their largely apolitical outlook aligns well with recuperative logic, but it also doubles as a political and rhetorical tactic in its own right.

"Blinkenlights" is a slang term for the illumination of choice in many hackerspaces. Its industrial origins would lie in the diagnostic LEDs built into mainframe computers and networking equipment such as modems. However, in hackerspaces, the same LEDs are arranged and programmed on purpose to achieve aesthetic effects. Therefore, a dialectical reversal is achieved. LEDs constitute some of the shortest and simplest feedback loops between machines and operators. In industrial environments, these serve to optimize the efficiency of production and, ultimately, increase the profit margin of capital. In hackerspaces, the same instruments serve the reproduction and recreation of workers, in order to prove Levy's adage from the first definition of hacker ethics: "You can create art and beauty with a computer" (1984, 35). Thus, Blinkenlights may be read as an allegory of workers' self-management.

ACCELERATORS

In 2005, the venture capitalist, hacker, and entrepreneur Paul Graham delivered a motivational speech on how to start a start-up to the Harvard Computer Society. At this point, his business-minded followers knew him as the developer of early web shops and as a technical author. Hackers and computer scientists read his influential and philosophical essays on what it means to be a hacker, programming techniques, and development methodologies (Graham 1993, 2004a, 2004b). The hacker legend insisted on three factors for successful start-ups: the technical prowess of the founders, which had to be high; the age of the participants, which had to be low; and their business knowledge, which had to be average. He mentioned in the talk that he had always wanted to be an angel investor since he became rich with his first start-up but had never got around to doing it in the previous seven years since he had become rich by selling his start-up to Yahoo! Inc. (Graham 2005).

In his account, Graham (2012) reflected on his own words while driving back home and decided on the same day to fund a large number of undergraduate students with small sums each to found their own companies that summer. The Summer Founders Program would only accept young undergraduates who were real hackers and would teach them about business in exchange for a 7 percent stake in the company. Successful companies would pitch their projects to venture capital firms later. The basic idea remained the same while the initiative grew into Y Combinator, the first instance of the "accelerator" genre of shared machine shops, which produced many globally known companies such as Airbnb and Reddit.

Accelerators can be found throughout the central economies of the world today. They can stand alone as profitable companies, attached to major players in the digital market segment such as Adobe Inc. (XD plug-in) or sponsored by local authorities from the European Union (EIC Accelerator Pilot) to city councils (see Bone et al. 2019). The original Y Combinator program soon moved from Boston to Mountain View, California—in the emblematic Silicon Valley. Paul Graham credits the move to his experience at Foo Camp that took place near Berkeley, California, an annual hacker convention organized by O'Reilly Media, the influential software development book publisher. Even though accelerators are a global phenomenon today, they still look to the Silicon Valley cluster of IT companies as a role model.

This may all sound very boring to readers who enjoyed the previous sections. It may not be at all evident that accelerators connect to the confrontational and anarchist politics of hacklabs or the do-it-yourself ethos of hackerspaces. We count accelerators among the list of shared machine shops because they explicitly draw upon the legitimacy and tropes of hacker culture, while targeting the same demographic group that could otherwise have ended up joining a hackerspace (i.e., graduate students during their time at technical university). The culture of hackers has been deconstructed and rebuilt from the ground up in order to create a climate chamber in which venture capital can thrive.

The narratives that hackers like to tell about themselves could, with only minor adjustments, be made to fit the new bill. As mentioned in the previous chapter on the open-source desktop 3D printer, engineering culture displays an ambivalent attitude toward the price mechanism. On the one hand, engineers decry the irrationality of allowing prices to dictate industrial activity, as this often leads to wasted expenditure and suboptimal output. On the other hand, price is generally accepted by engineers as a neutral gauge of the cost-efficiency, by which is understood superiority, of an engineering solution. On balance, the engineering profession has been more vexed about bureaucracy than markets. Catering to this disposition of engineering students was a small task. Hence, the market can be presented as a polling station for consumer satisfaction and a mechanism by which technology is democratized. The one thing that prevents consumers from accessing better products is incumbent interests that rely on monopoly practices and crony politics in order to stay on top. From this narrative, it follows that the role of the good guy is played by small entrepreneurs with big ideas who want to cut into the relevant market segment. Young engineering graduates with seed money can thus pretend to have a share in the rebellious and outlaw imagery of the computer underground.

The freedom allowed by seed money from Y Combinator is similarly ambiguous in terms of property relations. The first call (Y Combinator 2012b) and the application form (Y Combinator 2012a) emphasize that what the initial investment allows is merely for participants to move into a rental apartment in Berkeley and buy food for themselves during the summer break. This is in line with Graham's theory that high-level programming languages and consumer-grade network connectivity have brought down the cost of establishing start-up companies in the lucrative and attractive IT sector. However, it is also in line with the tendency of capital to provide, barely, for the reproduction of workers, relegating social mobility to the realm of possibilities, where it serves to justify the social domination of capitalists over the rest of society. In practice, the few examples of "unicorns"—unusually successful IT start-ups—produced by Y Combinator and similar accelerators legitimize their wider economic functions. The latter might boil down to launching into and maintaining excess working in orbit above the value-producing sections of the market economy, so that they neither rely on unemployment subsidies nor contribute to overproduction that would bring inflation.

The entrepreneur-hacker-workers of Y Combinator are granted the freedom to work whenever and wherever they want, yet are guided through the modulation of their life conditions to converge on the aforementioned apartments in Berkeley. Graham celebrates the rigor and flexibility of young graduate students for enabling them to work hard for the success of their own company in a self-directed way (Graham 2005). What he describes is easily recognized in Marxist terms as the formal subsumption of self-motivated labor. Accelerators bring into the service of capital the methodology of collaboration once developed by hacklabs and hackerspaces, to reduce institutional oversight—but in the hope of increasing productivity.

Y Combinator's office in Mountain View, California, situates the venture at the heart of the Silicon Valley capital, as part and parcel of a booming economic miracle. The dilapidated postindustrial buildings where hacklabs flourished are nowhere to be seen—neither are accelerators hosted in neighborhoods in the process of revitalization where hackerspaces once found cheap rents. In fact, the main opportunity offered to participants by Y Combinator is to insert themselves into the networks that constitute the industrial cluster of Silicon Valley.

In line with the formal subsumption of the legacy of hacklabs and hackerspaces, the Y Combinator company headquarters where events of the "start-up school" take place feature a few distinct—but easy to install—markers to show that it is not your boring average office. Clean orange walls mimic the color of Y Combinator's logo and enable audiences to identify the location of photos of participants taken on the premises. In contrast, hacklabs made no pretense of decoration, while hackerspaces fall at the other end of the spectrum with their elaborate Blinkenlights displays. The recycled infrastructures of squats are difficult to keep clean, and in hackerspaces cleaning is left to the (predominantly male) membership—but accelerators hire staff to clean and change failing light bulbs.

Accelerators are lit well and evenly, like any other office space or production plant. Lights are regularly changed by maintenance workers, who are responsible for stabilizing patterns of illumination. Illumination has always been a key consideration in the layout of the factory. It is essential for workers to be visible so that managers can oversee them, and visibility of the machines is essential for workers to operate them with maximum productivity. Lights go on and off at once in sizable sections of the space, in the same rhythm as the cycles of the eight-hour working day and the movement of the sun around the earth. Accelerators borrowed freely from the methodologies and cultural tropes of earlier generations of shared machine shops, but only those that expedited entrepreneurial ends. Blackouts and energy shortages were commonplace in the original shared machine shops. They contributed to the ambiance of the place, while reinforcing at an aesthetic level a shared opposition to living inside an office cubicle. Understandably, this was not a feature of hacklabs and hackerspaces that the designers of accelerators wanted to emulate.

FUNCTIONAL AUTONOMY ADVANCED AND RELINQUISHED

Earlier in this book, we identified three pillars of functional autonomy as key for enabling hackers to articulate critical perspectives on science and technology and, concurrently, for them to detect and resist recuperation attempts. These pillars are technical expertise, shared values, and historical memory. Shared machine shops are privileged sites for the articulation of technical expertise. They provide a social milieu within which technical expertise can circulate and develop at arm's length from the predominant institution in society where such knowledge is reproduced: the engineering curriculum. Equally important, shared machine shops preserve and expand the material infrastructure upon which the critical engineering practices of hackers depend.

Shared machine shops serve many more purposes than just facilitating the learning process and providing its members with technical skills. Informal socializing, gossiping, partying, and so on strengthen the ties of solidarity and build a cohesive identity among the participants, both within the local hackerspace and in relation to hacker culture at large. In the process, the hackers adopt a shared value system. The social aspect is most clearly forthcoming in the early genres of shared machine shops, hacklabs and hackerspaces, but also holds true for makerspaces and FabLabs. Even in the most recuperated manifestation of shared machine shops (i.e., the accelerators), there are social and ideological components in addition to the instrumental purpose of the setting. Participants understand themselves and represent their mission in opposition to what they identify as mainstream narratives about technological—and, therefore, political—exclusivity. The ambition of democratizing technology is embraced by all the genres of shared machine shops and constitutes a minimum of shared values.

In our assessment, the weakest link in the constitution of shared machine shops is the transmission of historical memory. Having a common understanding of past events and their significance for one's own collective existence in the future is indispensable if hackers are to maintain their functional autonomy over the long haul. We argue that the decisive moment of memory loss occurred in connection with the transition from hacklabs to hackerspaces. Soon after the latter movement was launched, it forgot, not to say repressed, the memory of its origins and, subsequently, the lessons that had been learned by the preceding generation of hackers. Thus, the hackerspace movement was ripe for future waves of recuperation, which took place in multiple ways and resulted in an explosion of depoliticized shared machine shop genres. Makerspaces, FabLabs, Tech Shops, and so on can be adequately described as successive steps toward the full consummation of a recuperative logic, the endpoint being the accelerators.

This observation can be made without denying the contrasting testimonies coming from participants. Hackers who contribute to "any" genre of shared machine shop may legitimately assert their experience of being part of something special: an empowering environment that challenges institutional control and managerial oversight. All genres of shared machine shops promote technical literacy and thus have the potential to contribute to expanded worker autonomy. However, when all the successive iterations of shared machine shops are placed next to one another in a longer time series, a trend becomes evident. The potential for generating critical engineering practices steadily shrinks, leaving a correspondingly larger space for capital to put the collective to work.

At the outset of this chapter, we made an adventurous methodological proposition with reference to satellite images of the earth, where brightness indicates areas of economic activity and boom times, while deprived areas and downturns in the business cycle are detectable in a dimming of the light emissions. Our spin on this popular-culture trope is that patterns of indoor lighting in shared machine shops indicate the extent to which the space in question has been integrated into the circuits of capital accumulation. The light signature of hacklabs corresponds to a stochastic pattern that stems from recycled infrastructure and self-maintained hardware. These material conditions in turn speak about the marginal position of hacklabs and house occupations vis-à-vis surrounding society. As for hackerspaces, the ambiance of those places is in large part created by illuminations programmed to perform eccentric geometric and temporal patterns. The Blinkenlight aesthetic is a display of workers' control. Diagnostic lights, originally used to supervise the production flow in the factory, have been repurposed for aesthetic pleasure and playfulness, announcing to the world the self-organizing splendor of hackers. Finally, the illumination fingerprint of accelerators is barely distinguishable from that of a conventional office space. The accelerator's light emissions ebb and flow with the same pulse as regular office hours.

Taken together, the visual narratives of illumination patterns from these three genres of shared machine shops tell the story of their gradual reintegration into capitalism. The shadowy abode of the hacklab testified to the social marginality of its denizens. In hackerspaces, its participants expressed their self-managed cultural distinctiveness by making a show out of the lights. The accelerator, finally, has blended with and become indistinguishable from its surroundings, even in terms of indoor lighting.

THE CONTEXT OF GENTRIFICATION
AND CAPITALIST RESTRUCTURING

In the following, we zoom out to the third time horizon in our analytical scheme, with the ambition of substantiating the claim previously made, that we can observe a tendency of development within the movement of shared machine shops toward the full consummation of the recuperative logic. This narrative is framed by two systematic ruptures in capitalism. As indicated in the section on the prehistory of shared machine shops, the story begins with the economic crisis of the 1970s, in the aftermath of which came deindustrialization and the outsourcing of industrial production. The story closes with the consolidation of urban regeneration programs, which began initially in the 1990s with the rhetoric about "creative cities," but picked up speed during the 2010s, this time under the banner of the "smart city." The movement of shared machine shops can be situated in between these temporal landmarks, the fall of the industrial city and the rise of a city of finance and media consumption.

Each genre of shared machine shops can be related to different conceptions of the city. This follows simply from the fact that physical spaces within a built environment will be subject to trends in city planning. Urban development policy is consequential for shared machine shops in more indirect ways as well. The social and economic conditions of the immediate vicinity will be reflected in what goes on inside those buildings, as suggested by hackerspaces sheltering Occupy activists and chess players. Above and beyond the empirical case for making a digression into gentrification processes, this reference lends support to two key theoretical claims in this chapter. Firstly, speculation in real estate demonstrates how value could be derived from secondary activities and interactions that took place outside of the formal wage relation and despite the actors having different goals in mind. Secondly, we can diagnose the progress of recuperation by reference to the different political agendas that the different genres of shared machine shops have enacted within urban regeneration plans.

Hacklabs emerged during the 1990s, for the most part concentrated within regions in southern Europe that had suffered from deindustrialization and capital flight. Following urbanologists Stephen Graham and Simon Marvin (2001), media theorists Plantin and others point to the retreat of modernist and infrastructural ambitions during this period as constitutive of the symbolic meanings and social formations of an early computer culture (2018, 299–300). A more palpable consequence of the economic downturn was an abundance of empty buildings in city centers. In combination with an underperforming urban policy environment, the stage was set for a movement of occupied social centers to blossom (López 2013).

In league with squatters, hackers filled the empty buildings, catered to unfulfilled social needs in their neighborhood communities, and exploited underdeveloped infrastructures that had been abandoned after the first wave of neoliberal urbanization. The annual hackmeetings, the hacklabs, and legendary events, such as Nocturnes and the actions of Escamot Espiral (detailed above), took place in occupied social centers. The squatting movement provided the material and cultural conditions for hacklabs to emerge. On the downside, as has become evident after this movement fell away, the autonomist milieu also set the historical limits for the expansion of a highly politicized form of hacking.

The instigation of hackerspaces, and much the same can be said about makerspaces and FabLabs too, was predicated upon the struggle for access to the city that had been waged by squatters and hackers during the previous years. In their confrontations with city halls, occupied social centers were adamant about their cultural and social contributions to their neighborhoods. This community service was, however, entangled with a political agenda: fierce opposition to neoliberal policies, advocacy for free spaces for artists and activists, and campaigns demanding affordable housing for local residents. In the same period, municipal councils courted Richard Florida's rhetoric about "creative cities" (2005). Under this auspice, artists and community organizers were instrumentalized for the purpose of revitalizing formerly neglected, inner-city quarters. Working-class neighborhoods and factory sites were refashioned and gentrified, in the first step by setting up artists' workshops and offices of community associations. Hackerspaces fitted the bill.

During a transitional period, hackerspaces had an easy time securing tenant contracts on favorable terms in central locations. The underlying agenda of the practice is suggested, for instance, by the name with which city administrators and entrepreneurial artists refer to such places. In the Netherlands, they are called a *broedplaats*—that is, a "breeding ground" or "hatching place." It is understood that the broedplaats is only a temporary arrangement, to be terminated when the neighborhood becomes ripe for more rentable investments. The founding decision of hackerspaces, in contradistinction to hacklabs, to self-organize within the confines of legal and institutional arrangements, makes them prone for enrollment into neoliberal urban policies. In some instances, this practice is directly pitted against the autonomous movement and, subsequently, against the hacklabs. A case in point is the situation of some hackerspaces in the Netherlands—such as Frack in Leeuwarden or Nurdspace in Wageningen—which were hosted in buildings managed by antisquatting companies. These companies offer precarious contracts for tenants in small parts of otherwise empty buildings in order to protect the property from squatters.

In the introduction to this book, we noted that recuperation often takes place behind the backs of participants. Something familiar acquires a new meaning that is diametrically opposed to what it once stood for, while the outward appearance of the thing remains the same as always. This is exactly what has happened when the soldering irons, the mandatory do-it-yourself slogans, and free software advocacy, which used to be part of the inventory of occupied social centers, resurface in a community space whose ultimate purpose, as stipulated in the tenant contract, is to protect and nurture the interests of real estate developers.

The policy discourse about "creative cities" has largely been supplanted by talk about "smart cities." Creativity was the watchword in the promotion of broadband connectivity a decade ago. The communication infrastructure was imagined as laying the foundations for unlimited and ubiquitous expansion in directions that could not be specified in advance by policy makers. In contrast, 5G networks are deployed under the auspices of smart cities, where the whole environment is designed from the ground up with specific business and administrative applications in mind, integrating everything from autonomous vehicles to crowd control. Where creative city evangelists rallied around the entrepreneurial, small-is-beautiful ethos as the vector of economic development, smart city lobbyists construct a rationalist, top-down narrative of technocratic control (Luque-Ayala and Marvin 2020).

Accelerators fit perfectly into this latest vision of the city. In doing so, however, their advocates mobilize the technological imaginaries, material practices, and tropes of hacker culture and previous generations of shared machine shops. The DIY imperative, which merged with direct action in hacklabs, and was reimagined as "do-ocratic governance" in hackerspaces, remerges as agile development practices in accelerators and similar organizations (Irani 2015b). The name "Y Combinator" alludes to a design pattern from functional programming, a programming paradigm that is strongly attached to the alternative engineering culture of hacking (the canonical reference is Abelson and Sussman 1996). While functional programming languages and techniques are seldom used in the industry, hackerspaces host functional programming meetups and cultivate this paradigm.[13] As mentioned above, Y Combinator founder Paul Graham is a notable functional programmer who has published on the topic (1993), but the legacy of hacker culture was quickly watered down in the hands of his many acolytes. Hackerspaces supplanted hacklabs while preserving some of their predecessors' political engagements. The recuperation of hackerspaces proceeded through structural conditions and neoliberal policy making that engulfed the affordances and foresight of individual members of those spaces. Accelerators, in contrast, recuperate hacker culture by design.

Gentrification is the common thread running through this narrative about genres of shared machine shops. Each successive wave has had to adapt to an urban landscape that was even more gentrified than the preceding ones. With each generational shift, the shared machine shops have allowed themselves to play the part of a vector for the same development trend. In support of this claim, we highlight a number of instances when local residents and community activists have resisted gentrification by targeting later generations of shared machine shops. In Barcelona in 2014, a FabLab promoted by the city council and specializing in experimentation with new materials was attacked and damaged shortly after it opened its doors to the public. The FabLab had displaced a soup kitchen in an until then abandoned warehouse where *chaterreros*, undocumented immigrants who get by on collecting and recycling metal, camped and worked (Ribera-Fumaz, September 26, 2014). In 2016, La Cantine FabLab in Nantes was set on fire. Nobody took responsibility, and the perpetrators were not found. However, the next year anarchists torched the local FabLab and media lab in Grenoble (Soufron 2017), claiming that, "City managers cater to start-ups greedy for money and the fashionably geek masses by opening FabLabs in trendy neighborhoods. These extremely diverse measures on the surface all aim to accelerate the general social acceptance and usage of the technologies of our sinister era" ("A Fablab Burned Down in France by Anarchists" 2017, para. 25; "Grenoble Technopole Apaisée?" 2016).

From these examples, it is plain to see that shared machine shops, which originated in autonomous movements and took part in contestations against neoliberal urbanization plans, have evolved to a point where antigentrification activists nowadays identify them as instruments of the same oppressive policies.

CONCLUSION

By situating our study of shared machine shops within a longer time horizon, we have put forward the argument that this movement owes its existence to industrial conflicts in the past, including such precursors as the Lucas Plan, socially useful production, collective resource approaches, and the science shop movement. Their predecessors were defeated in the economic downturn of the 1970s, in the wake of which came deindustrialization, factory automation, and the outsourcing of production. Thus, a new cycle of capital accumulation was inaugurated, sometimes referred to in the literature as "post-Fordism" or alternatively, the term that we are using here, as "informational capitalism." Concurrent with the restructuring of the economic system, the institutional means by which the Fordist mass worker had contested the reign of capital (i.e., trade unionism and electoral politics) entered into decline. It also spelled the end for the experiments in bottom-up decision making in workplaces and workers' participation in design processes that had taken place on the fringes of those institutions. When such initiatives resurfaced anew, of which shared machine shops are an example, the terrain of struggle had shifted toward autonomist practices and tactics.

In the world of hackers, the institutional framing of the struggle against managerial authority over the labor process has fallen away, ceding to struggles over the self-management of the labor process. Putting our findings in terms of Cleaver's conceptual work, shared machine shops exemplify a circuit of struggle for the self-management of labor outside the circuits of capital accumulation, by means of creating autonomous social relations. Hackers contest capital's prerogatives over scientific research and technological development by organizing a physical infrastructure and a cultural context within which alternative pathways in development can be pursued. In this analysis, shared machine shops are a form of political intervention that has more in common with direct action, occupations, and wildcat strikes than is suggested by its prehistory in union negotiations, electoral politics, and participatory design exercises. Critical engineering practices contain the potential, however slight, for production in society to be liberated from the confines of regimented working hours, managerial hierarchies, and occupational identities.

However, as we have documented at length in the above pages, successive iterations of shared machine shops advanced along the familiar trajectory of critique and recuperation. The confrontational and militant practices of hacklabs were marginalized and superseded by the more consensual and pragmatic outlook of hackerspaces. This set the stage, passing through a number of intermediate steps, such as makerspaces, FabLabs, incubators, and so on, for the consummation of the recuperative logic in accelerators. In spite of the activist origins of this movement and the nonalignment of hackers with formal institutions and occupational structures, latter-day genres of shared machine shops are instruments of urban gentrification (real estate speculation), industrial hegemony (consent to work), and capital accumulation (open innovation). At an aesthetic level, the same tendency toward the full integration of the computer underground into mainstream society and everyday work routines can be read from changes in the illumination footprint of shared machine shops.

Saying this is not to detract at all from the strategic importance of shared machine shops, which in principle—and, occasionally, in practice—nurse critical engineering practices and inform disputes over research and innovation policies. Assisting the public's involvement in comprehending the political underpinnings of science and technology is as urgent as it ever was. Ultimately, the point of our reconstruction of the history of shared machine shops is to restore the memory of the militant origins of this movement, so that the lessons learned by past generations of hackers can give direction to cycles of struggle yet to come.

6

INTERNET RELAY CHAT
A TIME MACHINE THAT STOOD
THE TEST OF TIME

THE PROMISE OF A TWO-SIDED COMMUNICATIONS
INFRASTRUCTURE: FROM RADIO TO IRC

In 1932, the radical playwright and theorist Bertolt Brecht famously called
for the democratization of radio. In doing so, he sketched the outlines
of coming struggles over communication infrastructure. The remarks he
made have as much bearing on present-day computer networks as they
had for radio technology back then: "Radio is one-sided when it should
be two. It is purely an apparatus for distribution, for mere sharing out. So
here is a positive suggestion: change this apparatus over from distribution
to communication. The radio would be the finest possible communication
apparatus in public life, a vast network of pipes. That is to say, it would be
if it knew how to receive as well as to transmit, how to let the listener speak
as well as hear" (Brecht 1964, 51).

In the following decades, experiments with building two-way communication networks using radio technology flourished. A milestone was the
decision in 1957 by the US Federal Communications Commission to designate the 27 MHz shortwave frequency range to citizens band (CB) radio.
Arguably, CB radio resembles Brecht's "vast network of pipes" in the sense
that the radio operators exchange messages with other users in the vicinity
without requiring stationary infrastructure.

XKCD 1782: Team Chat (Webcomic, Randall Munroe, 2017, https://xkcd.com/1782/. This work is licensed under a Creative Commons Attribution-NonCommercial 2.5 License. (Used with the permission of the author.)

The CB radio network played some role in social conflicts in the past. A case in point is a strike in 1974 organized by truck drivers in the United States. A labor ethnographer who documented the events reported that "several [interviewed truck drivers] suggested that the citizen band radio networks played a role in planning, publicizing, and enforcing the nation-wide shutdowns" (Bisanz 1977, 63). A local newspaper introduced a vocal participant: "J. W. Edwards, who gained some fame using his radio code name 'River Rat' during the nation-wide truckers strike a year ago, says truckers still have no strong voice to press their demands" ("'River Rat' Says Truckers Forgotten", 1975). Eventually, "the citizen band radio became a cross-country broadcasting station through which a communication network was developed among truckers" ("'River Rat' Says Truckers Forgotten", 1975). Thus, the alternative usages of radio technology enacted a critique of one-way, broadcasting-only media, in keeping with the analysis first offered by Brecht.

In her study of the culture of amateur and community radio, Christina Dunbar-Hester contrasts this setting with hacker culture and the ongoing struggles over digital information infrastructures. Of particular note is the mourning she reports among many amateur radio activists over the transition from analog radio technology to online radio. The activists anticipated an erosion of shared values and historical memory within their subculture due to the changing media landscape (Dunbar-Hester 2014). Their sense of loss resonates strongly with the arguments we have put forward in relation to generational shifts within hacker culture. In this chapter, we trace a similar attachment among tech-savvy computer users to an—allegedly obsolete—communications infrastructure, in which they have invested aesthetic and ethical values more dear to them than the advanced features offered by novel chat devices. In their stubborn refusal to switch from their antiquated chat device to the hegemonic social media platforms, hackers uphold a reservoir of memory of the utopian promise once embodied in the internet.

The democratic infrastructure of radio technology served as a reference point when political imaginaries were being formulated in connection to the nascent digital media during the 1970s (Enzensberger 1870; Túry 2015; Wyver 1995). Initially, digital chat protocols were designed with an eye on the world of amateur radio.[1] Internet Relay Chat (IRC) was set up in 1988, and until this day, some of its descendants remain operational, including the first server, irc://tolsun.oulu.fi (Frechette and Rose 2007). The terminology surrounding IRC bears witness to its historical debt to amateur radio. In contrast to many corporate-owned chat devices, where the term "chat rooms" is frequently used, multiuser chat on IRC is conducted in "channels." These revolve around an advertised "topic" that is visible at the top of the conversation at all times, and popular chat applications identify the channels by numbers, reminiscent of radio channels on CB and VHF bands.[2] Users have "handles" or "nicks," rather than "user names" or "profiles." Handles in the CB radio vocabulary identified individual operators and were chosen by the operators themselves. Furthermore, the moderators of IRC channels are called "operators," another reference to radio practices. The "relay" in the protocol's name refers to radio stations acting as repeaters of signals.

In saying that IRC "relays" chat messages, this is not merely to pay tribute to the history of radio in name alone. It reflects an underlying design choice, which is loaded with political significance. The servers broadcast messages to users who are currently tuned to a channel, without that data being stored on the server. IRC is conceived as a streaming protocol, in contrast to commercial chat devices that tend to be built according to the store-and-forward model. In the store-and-forward model, messages remain on the server, providing a complete record of past conversations.[3] Streaming is cheaper and faster, but, more importantly, messages disappear once they have passed through the server. It is easy to see how crippled such a design solution must appear from the vantage point of a business model centered on mining data and selling the information to third-party customers.

Many design features have been carried over in the transition from analogue radio to digital chat. IRC servers federate into "IRC networks," where several servers, often operated by different administrators, syndicate the same messages. Users can choose to connect to one or another server, but all the servers within a single network provide the same services and serve the same content. Similarly to the social dynamics of CB and amateur radio, networks reflect geographical and language barriers, as well as the opinions of users about the appropriate technical settings and social norms. Crucially, no coercive mechanisms are embedded in the protocol, whereby dissident users could be prevented from setting up their own networks over which they define their own rules.

The IRC case study plays a key role in our overall theoretical argument about recuperation. In contrast to the other empirical chapters, it displays the successful resistance of hackers against the recuperative logic of informational capitalism. Of course, this is a provisional statement that must be continuously reassessed. Our argument is presented in three parts. Firstly, we trace the trajectory of IRC use through three decades of internet history. Secondly, we survey the contemporary use of IRC as the backstage communication medium of peer production communities. Thirdly, we draw some general conclusions about what can be said to have been lost and what is carried over in cycles of critique and recuperation. This we do by reconnecting the precedent of two-way communication that amateur radio set for the later development of IRC networks.

THE EARLY INTERNET ERA (1985–1995): NONCOMMERCIAL NETWORKING

IRC was born around the same time as the World Wide Web (1988–1989). The chat function evolved from services on the bulletin board systems (BBSs), whose history has been documented on film by Jason Scott (2005). BBSs constitute the missing link between analog and digital media. They ran on digital computers, but users dialed in through analog phone net-works. Once logged in, the BBS presented a custom-built menu with options such as a discussion forum and a file download area. Chat services were sometimes implemented, but their usefulness was limited, since each user clogged up an entire phone line while logged in to the system. Unsur-prisingly, the first IRC developers and administrators were BBS operators (Frechette and Rose 2007). Kevin Driscoll (2016) argues that BBSs were the site of emergence for the features and sociality associated with contem-porary social media platforms. However, BBSs differed from social media platforms in that they were designed and operated by amateurs, relying largely on hobby hardware and domestic phone lines. IRC servers ran on university hardware, yet they were also designed and operated as the pet projects of their operators. IRC evolved as a side project without research funding, unlike similar developments in internet protocols, such as TCP/IP in the United States.

Jarkko Oikarinen designed the first version of the program and oper-ated the first IRC server. It grew from his OuluBox BBS while he was a system administrator in 1988 at the Department of Information Process-ing Science at the University of Oulu, Finland. He remembers that "they didn't have much for me to do" and that administering the university server "didn't take all the time" (Frechette and Rose 2007, para. 6). When more resources were needed, Oikarinen "asked some friends of mine to start running irc [sic] servers in south Finland, mainly in Tampere Univer-sity of Technology and Helsinki University of Technology" (Frechette and Rose 2007, para. 10). After "internet connections to states started working" (Frechette and Rose 2007, para. 12), he goes on to mention universities in the United States in his recollection, coupled with the names of individuals who took initiatives. There is no reference to any official framework, man-date, or authorization: the development and deployment of IRC servers was a matter between university technicians. Thus, the culture of BBSs was brought from home computers to universities by system administra-tors who had direct access to untapped resources.

There is an inherent ambiguity between access and control that runs as a red thread through the history of IRC. In most cases, the communication networks have been managed by system administrators with a broad man-date from their university administrations. IRC represents an interesting case of workers' self-management of fixed capital that they can technically access, but not legally control. It is central to our argument that the chat medium would not even have seen the light of day in the absence of the functional autonomy enjoyed by tech workers. This autonomy stems from a combination of technological complexity and generous research funding. Therefore, it is worth tracing the debate about the functional autonomy of engineers to its political origins in struggles at the point of production. Workers' control over the means of production have been extensively theorized in relation to occupied factories and social centers in the classic case of the Autonomia movement in 1970s Italy (Pansa 2007).

Steven Wright, who wrote a documentary history of Autonomia, reports on an early debate in the 1950s over the class position of "technicians" (2002, 101–106). These technicians were involved in both the conception and the execution of production (Wright 2002, 103), so the functions of management and workers overlapped in their case. In a decisive intervention in the debate, Bologna and Ciafaloni argued that the technicians' position within the division of labor was a potential advantage for developing working-class power in the factory. They pointed out that management could only control technicians insofar as they shared their technical expertise (Bologna and Ciafaloni 1965). This observation about the relationship between workers' power and technical expertise illuminates an intersection between the concerns of autonomist Marxism and labor process theory. It is from a proponent of the latter perspective, David Montgomery, that we have derived the notion of "functional autonomy," which is precisely what the Italian technicians possessed thanks to their intimate familiarity with the details of the production process.

In the thinking of the Autonomia movement, the appropriation of resources owned by the employer, through embezzlement, sabotage, refusal to work, and so on, was designated as a key terrain of struggle against capital. Similarly, we can also attribute such an aspect to the emergence of IRC, although typically it was not framed in such confrontational terms at the time. The consensual outlook owes its existence, firstly, to the privileged academic environment, which gave workers more leeway than in the factory. Secondly, the whole field of computing and networking was under development, and so, in the early days, initiatives and experimentation by the technicians were typically encouraged. Thirdly, as we will stress continuously throughout this chapter, the expenditure and resources required for IRC provision were low scale, allowing it to fly under the radar. All in all, the university accommodated a degree of workers' control—detached from property ownership—by allowing technicians to decide on the utilization of computing power and network capacity. In the theoretical chapter on critique and recuperation, we called this process, whereby workers take effective control over resources, "communization," as the antonym of privatization.

The development of IRC networks never benefited from the large-scale grants like the one that, for instance, the BITNET network secured from the Rockefeller Foundation (Grier and Campbell 2000). Many IRC administrators resented the institutional authority that follows from an official mandate in an organization and hence preferred to keep their work with IRC to the side of regular work tasks. This decision kept institutional bureaucracy, managerial control, and the legally imposed policing of content safely at bay. The increasing complexity of information technology provided cover for workers to organize production independently of managerial supervision. IRC emerged from the happy marriage of an overinvestment in computer departments and an underexploitation of the technicians working in those places.

In the absence of managerial control over the design process, users figured out ways to implement a technical system that would further diffuse authority across the network and distribute responsibilities to its users, including operating features and carrying out maintenance. Design decisions and resource allocations were made directly by administrators. Content was moderated by operators drawn from the pool of the most active users. Administrators could be found in a special channel traditionally called #ircd (for IRC daemon, the server component of the IRC infrastructure), and operators could be recognized by an @ sign in front of their handles (such as @maxigas). The power of these individuals stemmed directly from their access to key parts of the infrastructure, while their good standing in the community depended on their technical expertise and social tact in wielding such powers. These are clear indicators of self-management and self-regulation.

Naturally, neither self-management nor self-regulation guarantee an absence of strife. The history of IRC is a history of splits, documented in primary and secondary sources such as Stenberg (2011) and Latzko-Toth (2013). A split is when an IRC network of federated servers fragments into two separate networks. Splits allow different technical configurations, and, as a corollary, different social norms, to be introduced into each network. A common source of contention resulting in a split is disagreement about the distribution of responsibilities between the users and administrators of a network. Certain functions, such as reserving handles for particular users, may be left unmanaged, or automated by the administrators using IRC bots that compensate for missing features of the IRC protocol itself.[4] For instance, two historical networks active to this day take different stances toward providing continuity in the existence and management of particular IRC channels. Freenode supports and advises registering channels with the ChanServ bot, which protects IRC channel operators, or at least the founder (called "owner"), from losing control of the channel. IRCnet, another major network, is perceived as a more free place, since it has no ChanServ bot, leaving users to their own devices to secure their channel. The point is that the proliferation of IRC networks was and is driven by disagreements about the politics of running these networks and maps the development of such political differences translated into network policies and configurations.

Thus, IRC served as a link between the BBS and WWW for communities in transition. In the rapidly changing media landscape of the early internet era, the simplicity, flexibility, and even limitations of the protocol were a distinct advantage over alternatives. These allowed IRC to establish itself as "the" chat device on the internet. New internet users found IRC already deployed on the infrastructure as the default chat solution. This is significant, because the population of internet users exploded during the 1990s. IRC benefited from "path dependency"—the persistence of deployed solutions even if better alternatives exist (David 1985)—in the subsequent era of the dot-com boom.

DOT-COM ERA (1995–2005): MARKET DIFFERENTIATION, COMPETITION

The streaming protocol design allowed IRC to scale. As the number of users grew exponentially, the required computing resources increased logarithmically. In other words, IRC servers could accommodate many more users without needing much more in the way of resources. In comparison, doubling the amount of email accounts on a mail server requires doubling the amount of storage space, so that resource requirements grow at the same rate as the number of users. Therefore, other services—from internet access through website hosting to email accounts—needed more resources in order to be rolled out to virtually all citizens in the advanced economies. In a significant departure from the norm of the early internet, the prevailing ideology of the dot-com era was that such large-scale services should be financed through developing a business model.

The commercial opportunities in the booming IT sector encouraged start-up companies to experiment with novel business models including free services and products funded from venture capital. The mood of the dot-com era is captured with eerie plasticity by Thomas Pynchon in his novel *Bleeding Edge* (2013). Most of the experiments disappeared when the stock prices went bust. The business models that survived either settled into a subscription-based delivery, as is still the case with internet service providers (ISPs) that sell internet access, or they were geared toward income from advertisements. The few so-called "unicorns"—start-up companies that came out at the top of the subsequent market consolidation—built digital platforms in the form of multisided markets: cross-subsidizing free services through gains from other aspects of the venture, as we will see later (Srnicek 2016). Irrespective of the precise details of the business models and innovations, the sudden inflow of investor money, followed by a consolidation of market power in the hands of a few conglomerates, fundamentally transformed the relation between users and the technology they were using. Users were turned into consumers, and formerly self-governed user communities became walled gardens of brand-loyal customers, guarded by intellectual property rights and digitally enforced restrictions.

IRC continued to operate throughout this period of commercialization. It was not uncommon for internet service providers to donate machinery and employers' time to the operation and maintenance of IRC networks. Below-the-counter servers were provided on a *pro bono* basis since workers—and in many cases, the managers themselves—were invested in the community (Driscoll 2016). This amounts to the communization of resources around the IRC protocol. It helped that very few resources were required to run IRC servers, which could easily be provisioned on the old model of embezzlement. To this day, the legal status of either the IRC network operators or the IRC users themselves is in limbo, leaving them in an ambiguous relationship to contractual remuneration and managerial hierarchies.

During this period, all kinds of service delivery were privatized under the influence of neoliberal dogmas. Public services and utilities came under the stewardship of private companies in what is sometimes referred to as the "platformization of infrastructures" (Plantin et al. 2018). The notion applies to public transportation and classic utilities such as heating, and the same logic expanded to the means of electronic communication when they had matured sufficiently. Up until the 1990s, internet access, email accounts, and domain names were developed and provided outside of market circulation. The market was the default mode, however, when those utilities were brought to the masses at the turn of the millennium. In the assessment of Plantin and others, the privatization and commodification contained in the platformization of infrastructures amounts to nothing less than the collapse of the "modern infrastructural ideal" (299–300). According to the modern infrastructural ideal, the establishment and development of nation states went hand in hand with laying down networks and grids of roads, electricity, heating, and telecommunications across a territory in order to establish national sovereignty. This association between modernity and the state was disentangled by neoliberalism.

IRC would scarcely fit the modern infrastructural ideal, and perhaps that contributed to it evading the processes of commodification and platformization. A number of more or less consistent factors account for this historical contingency, which come together in the aforementioned ambiguity of IRC as a "privately" operated "public" service. Privatization targeted public services, justified by the idea that the price signal is superior in delivering the optimal allocation of resources when compared to public administration (expressed in Hayek 1948). While IRC resources were not allocated based on the price signal, they were also not publicly administered. As theorized above, IRC ran on a mixture of volunteer labor and fixed capital belonging to a workplace. Therefore, the classic criticism advanced against utilities by neoliberals—that public administration is inefficient for delivering services—did not apply to IRC networks.

Notwithstanding the confusion surrounding the proper place of IRC within public and private institutions, such conceptual ambiguity would not have saved it from a neoliberal restructuring if IRC had been considered a significant cost or a potential source of profit. Hence, we stress the lean resource requirements of IRC, which can be attributed to the streaming protocol design, as the final reason for it withstanding the onset of recuperative processes during the dot-com era. The streaming protocol design allowed IRC networks to accommodate an exploding number of new users without submitting to market pressures that would have transformed the service from a utility into a product—or in other words, from an infrastructure into a platform.

A direct attempt to recuperate IRC was made by Microsoft in 1996. An IRC client was included in the popular Windows operating system (Kurlander, Skelly, and Salesin 1996). The ramifications of this relate to the monopoly position of the firm over desktop operating systems at the time. The default option in the client was to connect to an IRC network run by Microsoft. The added feature would thus bring the majority of internet users under the wing of the company. This was one of several attempts by the company to leverage its market power over desktop computers into a vast user base in networked services. The company had an ambitious plan to bring IRC up to date with the multimedia expectations of the dot-com era.

The IRC client software was developed by the company's artificial intelligence research unit. They designed a graphical interface for the client that drew cartoons on the screen that illustrated the written conversation in real time, based on Natural Language Processing algorithms (Latzko-Toth 2010). Animal characters impersonated the users, including their facial expressions driven by sentiment analysis, bringing the smiley—as emoticons or emojis were known at the time—to the next level. The Comic Chat IRC interface was never popular with users, who found it both cumbersome and bizarre. The only artifact that went down in history from the whole enterprise was the Comic Sans font, which is still the laughing stock of internet users. Independent IRC networks continued to thrive. Microsoft never figured out how to make money from the largest online chat phenomenon of the time.

Many typical use cases of the internet were concentrated on IRC in this middle period in the history of the protocol, such as dating, file sharing, and staying in touch with friends, as well as topical discussion forums, dedicated to everything from soccer to controlled substances. From this milieu, new terms emerged (such as ASL for "age, sex, location;" see Chester and Paine 2007, 228) in connection with new features (such as DCC for file transfer over IRC) to better accommodate these widely popular applications. Over the course of a decade, specialized platforms were spun off to serve each of these variegated use cases. As a result, users gradually migrated away from IRC to commercial services built on different media technologies, and a diverse market of online content providers dealing in user-generated content was born.

For instance, the mainstream of electronic dating was initially located on IRC, then on commercial websites on the World Wide Web, and most recently on dating apps like Grindr and Tinder. Topical conversations, on the other hand, migrated to social networks such as Facebook and later Twitter. The latter could be—and has been (Blussé 2013)—discussed as a successful recuperation of IRC, since some of the social dynamics (for instance, flame wars and cancel culture), as well as elements of the syntax and semantics of the interface (the @ sign to identify users and the hash mark to identify topics), originated in IRC. However, Twitter is just the most indicative example of a general trend toward the diversification and commodification of online services, during which IRC went from unchallenged hegemony to a fringe medium.

While entertainment dominated IRC use, there were many other significant developments. Citizen journalism was one of them, with the following examples documented in an archive collected and published by David Barberi (1999). Ordinary people reported on their everyday experiences during the first Gulf War (1990–1991). Dissidents communicated on IRC during a coup d'état against reform-minded Russian president Gorbachev in 1991, since radio, television, and phones were under blockade. Activists organized mutual aid during the siege of Sarajevo (1992–1996). These were the first instances when IRC fulfilled the promise of democratizing media through two-way communication professed by Brecht in 1932.[5] Finally, IRC and mailing lists provided the communications backbone in the development of free and open-source software, which took off at the end of the 1990s. While mailing lists facilitate structured and reasoned public discourse, IRC channels allow informal conversation, mixing technical development with friendly socialization. Many free software projects grew on IRC channels. Collaborative development was one of the first use cases of the protocol and remains indispensable to IRC culture today.

The state of user expertise and the media landscape accommodated the limitations of the streaming protocol design. Users owned personal computers stationed at their home or workplace, and these personal computers were increasingly connected to the internet through flat-rate ISDN lines. Countries developed broadband strategies that aimed to increase the number of households connected to the internet on a flat rate. While the strategies themselves emphasized the connection speed, the decisive factor for IRC adoption was the increased stability and affordable price of the connection, since text-only chatting consumes negligible bandwidth. The average internet user could install the mIRC client software on their desktop computer, join a channel, and sometimes even keep the computer running for a long time in order to have a complete record of the conversation.

During the previous epoch of the early internet, IRC blended in with the infrastructural landscape of user-facing services. As the dot-com era progressed, the widespread privatization of most other services turned IRC into an exception to the rule. Its now exceptional position meant that the IRC medium took on strong associations with the values and norms of the early internet, preserving the historical memory of users' self-management practices and the technical expertise of its operators. These three factors—shared values, historical memory, and technical expertise—we referred to in chapter 2 as the three pillars of functional autonomy. We argued that they are indispensable for the development and reproduction of hackers' functional autonomy. The lean protocol design behind IRC infrastructures stemmed from these factors and helped to sustain them in the long term. Indeed, users continued to operate IRC networks and serve diverse purposes throughout the dot-com era without either the infrastructure or the community ever fully integrating into the capitalist system.

SOCIAL MEDIA ERA (2005–2015): MARKET CONSOLIDATION, PLATFORM MONOPOLIES

At the turn of the millennium, the promise of the internet becoming a two-way communication medium looked feasible on a grand scale. Silicon Valley firms embraced the popular demand for a democratization of media as a justification for their ascendance (Liu 2004; Barbrook 2007). Capital sought to satisfy those demands in a way that concurrently undermined the original rationale for the critique to be raised in the first place. Brecht envisioned a network of pipes for two-way communication in the hope that it would encourage self-organization among the workers and bring on the transition to communist revolution. Silicon Valley capital turned the internet into a form of mass media, even if an interactive and participatory one, thus averting such a scenario from unfolding.

Social media became the dominant form of online service delivery, thus providing the groundwork for surveillance capitalism to emerge as the hegemonic form of capital accumulation (Zuboff 2015). This was the crisis that brought the dot-com era to an end. In the social media era, market consolidation set in, so that the most popular services came under the auspices of one among a small handful of global conglomerates. This process has been extensively analyzed by Srnicek (2016, chap. 3). In parallel, ubiquitous network connectivity through smart devices resulted in a shift in the social status of internet access. Internet access became available and affordable to the overwhelming majority of the population in societies located at the center and half periphery of the world-system. The masses arrived in online spaces, which corresponded with a sharp drop in the technical expertise of the average user.

The fast-moving media landscape practically left IRC behind. Users and techniques had to bridge the historical gap between the ancient protocol design and the contemporary technological reality, which increased the necessary technical expertise for using the protocol. This presented a novel situation. As we have emphasized in the previous two sections, so far the limitations of the protocol had proved to be an advantage for adoption, be it for different reasons. In the early days of the internet, ease of implementation and flexibility of use meant that IRC could keep up with the rapid innovation that characterized the media landscape of the 1990s.

During the subsequent dot-com era, simplicity allowed IRC to scale with the exploding population of the internet, without the need for a business model that could have led to its commercialization and eventual integration into the capitalist system. Meanwhile, its flexibility could accommodate a wide range of popular use cases, as diverse audiences became internet users. However, in the present era of social media, these characteristics provide few advantages.

Market concentration means that competing chat devices have good engineering teams and a dedicated community behind them. The corporations that survived the dot-com era learned how to develop effective sales tactics for positioning their digital products, and how to support their market adoption by incorporating user feedback loops. Simplicity is an advantage only as long as resources such as programming hours and CPU cycles are scarce. However, the consolidated media monopolies are immensely profitable, and hence they can throw an overwhelming number of capable engineers at a problem, as well as dedicate enough resources to tackle technical complexity.

The very essence of what is understood as "flexibility" in engineering terms has changed in tandem with the contemporary media landscape. Flexibility now refers to compatibility with social media platforms, which is another way of saying integration with them (Helmond 2015). The point that users can adapt IRC to a variety of purposes, including integration with social media platforms, is moot when most actual development takes place on top of those platforms (Helmond, Nieborg, and Vlist 2019). Platformization means that monopolies obscure, then replace, the media landscape itself, and that landscape is decidedly hostile to the engineering ethos and community values associated with IRC.

In response to this new situation, IRC users go to great lengths to bring their medium of choice up to date with contemporary usability standards, while striving to avoid being absorbed by the platform monopolies and the secondary effects of platformization. It is necessary to examine these technical measures in some depth, so that we can gauge their theoretical import. Essentially, users establish continuity in their presence on chat channels by running their own servers, where complete logs of conversations accumulate, until they can read them out from a laptop or smartphone running an ordinary IRC client.

Expert users access IRC through their setup. The word "setup" suggests the image of a drum kit rather than a tool such as a singular hammer. Individual users of other chat devices experience a single piece of software tightly coupled with a single service, but a personal IRC setup is instead engaged as an infrastructure in its own right.

Heavy IRC users rely on helper applications that maintain a connection to the IRC servers with minimal interruption. A bouncer is an intermediate program (middleware) that connects to an IRC channel instead of the user and listens to the conversation in their name or relays their words to the channel. The bouncer logs everything that happens and replays those logs to the user when they connect to it with their preferred IRC client. Helper programs such as bouncers compensate for missing backup logs by being as persistent as the IRC networks to which they connect, thanks to the fact that both run on servers.

Using IRC through a bouncer readily lends itself to multidevice use cases. Multidevice users participate in IRC through different devices, usually a mobile phone and a personal computer. Multiple IRC clients can connect to the same bouncer, while the bouncer is logged in to a channel only once. As the example of ubiquitous computing shows, contemporary IRC users need technical expertise to compensate for the tension resulting from interoperability issues between the ancient IRC protocol design and the contemporary media landscape.

The business model of social media platforms is incompatible with the implementation of chat services as a streaming protocol. Surveillance capitalism is based on storing and analyzing the log files that represent user activity. Even if users keep a log of their own activities using a personal bouncer or another contraption, this is not aggregated in a central location where it could be monetized by the IRC network. In the absence of logs, it is difficult for the service provider to capture and monetize user-generated content. Technically speaking, IRC is resistant to recuperation, in part because it puts up a high barrier to entry for users, and in part because its design hinders the implementation of prevailing business practices for monetization.

ANOTHER GOLDEN AGE OF CHAT? (2015–)

While IRC lost its leading role as the de facto standard chat device on the internet, chat itself has enjoyed a resurgence in popularity. Most smartphone users nowadays have several different chat applications installed on their device. Despite the hype around virtual reality, augmented reality, and other similar types of rich multimedia experiences, plain text chat is the most popular form of communication on mobile platforms.

Notably, the very few companies that could break the monopoly of the Big Five (GAFAM: Google, Apple, Facebook, Amazon, Microsoft) on the digital communication market are all chat providers, such as WhatsApp, Telegram, and Signal. They advertise their security credentials as a core element of their business models and distinguish themselves rhetorically by opposing the surveillance capitalist model of accumulation. However, both Telegram and WhatsApp have recently been bought by Microsoft, while Signal is the smallest of the lot, struggling to find a sustainable business model. It is an open question whether companies that specialize in chat products will be able to disrupt the incumbents, or whether they will be swallowed up in the next wave of market consolidation.

In the meantime, the same media monopolies market their own smart personal assistants, aiming to translate their products into the chat format where IRC pioneered automation. This applies to dominant and growing platforms such as WeChat in Asia, but also to the leading Western firms mentioned previously. Examples abound, but Apple's Siri may be representative of this trend. IRC bots are an obvious antecedent of these smart personal assistants. They have complemented and extended the basic functionalities offered by the IRC protocol since the very early days of IRC networks. They can seamlessly blend the informality of a conversation with the sealing of business transactions and the delivery of information services. Chat as a form of online service delivery seems to offer a significant usability advantage over form-based web shops and button-based app interfaces, since chat integrates frictionlessly into the life worlds of users, imitating quotidian sociality.

Due to these developments, an increasing proportion of social relations is mediated by chat. Even in the sphere of organizing social movements, discussions and conspiracies have moved from public social networking sites on the World Wide Web to invitation-only chat channels, especially so in the more repressive environments, including the rapidly developing economies of Asia (Rogers 2020). There may be a window of opportunity here for self-managed, trusted services such as IRC, except that, in the current proposals for chat devices, encryption replaces oblivion as the last line of defense against surveillance and the exploitation of metadata.

CONTEMPORARY IRC USE

IRC is often considered to be a relic of the past with a dwindling user base. In contemporary scholarship, IRC is recognized for its historical importance as an intermediate link in the passing from bulletin board systems to social media monopolies such as Twitter (Hogan and Quan-Haase 2010; Rogers 2013; Dahlberg 2015; Nagel 2017). Although scholars acknowledge the continued existence of IRC in the contemporary media landscape, they do not consider that it contributes much to that landscape at present (van Doorn 2011; Leistert 2016). In contrast, we argue that the new media outlets of today would be unrecognizable without the sustained contribution of peer production communities that organize their collaborative efforts over IRC.

Peer production is a mode of production based on collective, collaborative, and networked labor, characterized by the self-selection of tasks and the role of nonmonetary remuneration for those tasks (Rigi 2013, 402; Benkler 2006, chap. 3). Peer production is sustained by the communities of developers and users that spring up from the collective endeavor (Shirky 2008, chap. 2). In chapter 2, we proposed a distinction between the community of peer producers and the two other configurations of digitally mediated production: crowds of users and clouds of click workers. Our categorization is based on the relative degree of functional autonomy that they enjoy (or not).

Even in the last decade, IRC has seen significant adoption by newly started peer production communities. Generations of new users rely on the protocol, proving that contemporary IRC use can be the result of a conscious decision between alternatives, rather than just a fact explained by path dependency or nostalgic motivations. At the same time, established peer production communities generally continue to use IRC for real-time communication, even after other types of users have migrated away.

Contemporary peer production communities use IRC as their backstage communication medium for collaboration and socializing. The fact that it is a backstage communication medium partly explains why it has received less attention than more visible media, even while the products of these communities have been extensively studied. Moreover, the simple fact of continued use blinds the innovation-centric perspective of new media scholarship to the contemporary significance of IRC (as argued in Maxigas and Latzko-Toth 2020). In order to correct this blind spot, we provide an overview of contemporary IRC use, establishing the claim that not only has it resisted recuperation by capital, but it has also found a strategic niche, which makes it a historically significant case, rather than a mere curiosity.

The obsolescence of IRC can be argued quantitatively through references to the number of total users on the major contemporary IRC networks, as well as the decline in the number of networks themselves.[6] Moreover, in qualitative terms it can be observed that neither the protocol itself nor the associated infrastructures that effectively make up IRC networks are developing at a significant pace any more. We are not debating any of these claims for the obsolescence of IRC. Instead, we argue that they need to be qualified, by looking at what is achieved through the medium and how it is actually used.

It is virtually impossible to understand—or even research—contemporary hacker cultures without understanding IRC. It brings together ephemerality and longevity in a way that is key for the ideological cohesion and practical coordination of a wide range of hacker projects. These specific features may have contributed to its continued use amongst specific groups that rely on both aspects to mobilize and organize largely volunteer labor.

The following sections describe contemporary IRC use by three different peer production communities that engage in the collaborative production of politics, hardware, and software. Anonymous hacktivists, hackerspace members, and free software developers are only some of the examples that could be cited. For instance, IRC also plays a significant role in the social life of Wikipedia, and it is an indispensable tool for cybercriminals as a command-and-control infrastructure. Therefore, the communities under consideration here are but a sample of hacker subcultures that rely on IRC.

ANONYMOUS HACKTIVISTS

Anonymous is a worldwide hacktivist movement that exploded in popularity in 2007 and continued to thrill the public imagination at least until the arrest of some key members in the last few years and the subsequent outgrowth of alt-right activism from the same milieu (Coleman 2014; Nagle 2017). The founding myth of Anonymous is inexorably tied up with image board culture, and specifically the 4chan image board with which participants had a turbulent relationship (Zeeuw 2019). While image board participants have always used IRC to discuss topics of interest, Anonymous took chatting to a new level, by focusing on organizing online or offline direct action directly through IRC (Dagdelen 2012). Introduced to mainstream audiences in an inverted crusade against the Church of Scientology (called Project Chanology, 2008), the group continued to wreak havoc, taking the Mastercard payment processing servers offline in revenge for their economic blockade against WikiLeaks (under the operation code name Payback is a Bitch, 2010), and their splinter group LulzSec went on to put the CEO of the lucrative defense contractor cybersecurity firm HBGary Federal out of business. Anonymous's ability to pull off these operations is often attributed to the fluidity of its organizational form: "Anonymous is the first internet-based super-consciousness. Anonymous is a group, in the sense that a flock of birds is a group. How do you know they're a group? Because they're traveling in the same direction. At any given moment, more birds could join, leave, peel off in another direction entirely" (Landers 2008).

As Coleman's 2014 online ethnography shows, while audience-facing IRC channels can easily feature thousands of users on a single channel at high tide, sometimes the threads of a conspiracy are woven on invitation-only backstage channels with only a close circle of participants. Both the cacophony of the unwashed masses on the open square and the whispered plotting in the dark corners are essential for the overall success of Anonymous. The amalgam of mass movements and competing avant-gardes characterizes the peer production of poignant political imaginaries and operations. The particular sociality that constitutes the core of Anonymous's appeal as a next-generation hacktivist movement builds on the real-time technological mediation specific to IRC as a chat device.

In a show of versatility for the protocol, Anonymous groups run their own IRC servers, and typically steer clear of the major public IRC networks. On the one hand, they are too edgy to be tolerated for long on major public IRC networks, since they inevitably bring undercover agents and state repression in their wake. On the other hand, they are too paranoid themselves to trust the operators of the major public IRC networks. As IRC is an open standard and its major implementations are published under free licenses, all it takes to set up an independent IRC network is access to a basic server computer and rudimentary system administration skills. Therefore, participants could learn the tricks of the trade on public networks, and when push comes to shove, they can set up their own as a launchpad for their campaigns. Expertise in operating and using IRC networks is transitive.

Participation, expertise, and infrastructures built up, combining IRC users into a veritable cybernetic invasion force that Anonymous habitually mobilized against its opponents (Sauter 2014). Distributed-denial-of-service (DDOS) attacks bring down target websites by flooding them with data packets originating from all corners of the internet. These packets are typically "sent" by a combination of human and nonhuman agents during Anonymous campaigns. On the one hand, humans logged in to an IRC channel switch on the dreaded zombie mode of the infamous Low Orbit Ion Cannon (LOIC) software installed on their computers. Zombie mode lets the IRC channel operators aim the LOIC software on the users' computers in concert and orchestrates concentrated attacks. On the other hand, more tech-savvy Anonymous participants use private botnets composed of herds of infected computers. The malicious software on the infected computers also logs in as nonhuman IRC users to secret IRC channels. Like LOIC participants, they are controlled by the channel operator to carry out their work—but without the knowledge or consent of their users.

It bears to be stressed that a tool such as IRC lends itself for a variety of uses and agendas—likewise with the swarming tactics that have emerged in tandem with this tool. The drift of the Anonymous movement from one end of the political spectrum to the opposite end illustrates this ambivalence. Anonymous made it to the headlines when it launched operations in support of the Occupy movement in the aftermath of the 2008 financial crisis. A few years later, alt-right activism spawned from the same forums, and the same tactics were deployed in support of Trump's election campaign in 2016 and the storming of the Capitol in 2021 (de Zeeuw et al. 2020). The political outlook of a peer production community is not determined by the question of public access to the tools. Although it is not a sufficient condition for political emancipation, having access to the tools is an indispensable stepping-stone toward the realization of such a goal. The story about the Anonymous movement demonstrates the continued relevance of IRC for coordinating collective action.

HACKERSPACE MEMBERS

Hackerspaces were extensively discussed in the previous chapter on shared machine shops. In this chapter, we are interested in the intersection between hackerspaces and IRC as part of the material infrastructure for peer production communities. Most of the objects produced in this setting are for personal, educational, recreational, or research purposes that never make their way into market circulation. Such expressions of technological creativity may be best seen as a byproduct of the specific sociality cultivated in the physical space. Quantitative surveys (such as Moilanen 2010) and ethnographic accounts (such as Davies 2017) confirm than the main output is not the gadgets per se, but a specific form of subjectivity.

Hackerspaces rely on a consistent set of digital tools to organize the life of the community. Virtually all hackerspaces have adopted a similar repertoire: a website for public relations, a wiki for project documentation, a mailing list for formal discussions, and an IRC channel for day-to-day operations, socialization, and coordination. The channel is registered with one of the major IRC networks, so that it is widely accessible to the general public, just like the hackerspace itself. Users can join the conversation irrespective of membership status or geographical location. However, just as in the actual hackerspace itself, where they are entitled to the key to the door, members typically get operator status: pretty much the only formal gatekeeping mechanism in the physical space and the IRC channel.

Bots bridge the on-site infrastructure of the hackerspace with its virtual lounge, the IRC channel. These software agents announce on the chat when a member opens or closes the door of the hackerspace. They may report other statuses thanks to networked sensors, such as the temperature of the beer in the fridge or the availability of the laser cutter in the machine shop (as in Hack42, in Arnhem, the Netherlands). Similarly, they remind the audience of upcoming events and keep various statistics such as the current size of the membership, or the financial status of the organization (as in H.A.C.K., in Budapest, Hungary). Channel participants may be able to trigger events in the hackerspace through issuing commands to bots, such as displaying messages on an LED matrix or flashing lights to draw attention to the IRC channel (as in Technologia Incognita, in Amsterdam, the Netherlands). Therefore, there is a strong continuity between the online and offline manifestations of the hackerspace, which is achieved through an augmented reality approach based on "smart" devices—and "old" social media.

FREE SOFTWARE DEVELOPERS

Since 1989—which saw the publication of the first free software license (the GPL) one year after IRC had been set up—free software has become a major force in the information technology industry, powering most servers on the internet and smartphones on mobile networks. Free software refers to programs whose license is intended to protect the rights of users, instead of the producers (Stallman 1993). Its advocates promote it as a universally accessible pool of shared resources—in the manner of arguments by Ostrom and others (1999) in relation to the commons. There are many differences between free licenses, which are the subject of lively discussions between hackers, lawyers, and scholars.[7] The Debian GNU/Linux operating system distribution is a prominent and eminent project in the field of free software. It is the subject of Gabriella Coleman's book-length ethnographic study of free software developers. She highlights the central role of IRC in the development of Debian: "Much of the work on Debian happens in an independent, parallel, distributed fashion through informal collaboration on IRC channels or mailing lists, where developers ask for and receive help" (Coleman 2012, 128).

While she refers specifically to Debian, all peer production communities, by definition, organize the labor process in an "independent, parallel, distributed fashion," whether we consider Wikipedia editors, hackerspace members, or Anonymous hacktivists. We argue that, today, IRC supports peer production communities through specific affordances, complemented by those of mailing lists.

Mailing lists are kept on-topic through the affordances of that medium and the gatekeeping work of moderators. In contrast, discussions on IRC are more off-key, broadly tuned, and freewheeling. Conversations on IRC bring a playful, informal element into collaborative production: "Prospective developers are encouraged to join mailing lists and IRC channels that provide the medium for technical as well as social communication" (Coleman 2012, 142).

A host of topical IRC channels are associated with the Debian project. These range from channels dedicated to supporting end users, through channels coordinating the developers of certain parts of the system, to "off-topic" channels explicitly dedicated to socialization. We mentioned that the use of IRC for file sharing or online dating has declined over time, but the same cannot be claimed for free software development.

The IRC networks embraced by free software developers have experienced a steady rise in user numbers. This tendency is best exemplified by the current Libera Chat network, whose staff, users, and servers trace their genealogy back to Freenode, and even earlier, to the Open Projects Network. These networks are officially dedicated to "peer-directed projects" (Mutton 2004, 2), a terminology that is intentionally inclusive of all forms of peer production, not just free software development. That said, the infrastructural underpinnings of any IRC network depend heavily on free software projects for their reliability and flexibility.

Developers have kept their IRC channels up to date with current developments in software engineering, mainly through adding bots that interface the protocol to push and pull data to and from various sources. Similarly to the way in which Anonymous participants synchronize their weapons, or hackerspace members learn about the status of their manufacturing machines, software developers learn about and interact with their software repositories using IRC bots. To this end, various automation and abstraction mechanisms have been introduced into mainstream software development practices in the last decade.

Firstly, distributed version control systems keep track of each change that developers make to the software. These highly complex systems can be useful for various reasons, but also make it hard to keep tabs on what is happening with the software overall. Bots are used to announce changes and produce statistics about the data in the version control system. These reports and metrics allow participants to better follow how the software is evolving.

Secondly, continuous integration infrastructures make sure that recent changes do not break compatibility with previous versions, that test suites safeguarding software quality are passed without errors, and that the various parts of the software's ecosystem remain synchronized. Continuous integration infrastructures run in the background without user intervention, alerting users if something goes wrong. Developers may learn about these programmatically detected mistakes through a bot relaying the error messages to an IRC channel.

Thirdly, Agile and its associated software development methodologies introduced a resurgent interest in organizing collaborative work, emphasizing lean, iterative, and empowering management methodologies. Without even trying to do justice to these new software development methods or philosophies, we note that so-called stand-up meetings are often part of the process. These flash meetings serve to expose current status and the critical stumbling blocks encountered by individual developers to the whole of the team. Virtual organizations often conduct such meetings over IRC.

The early adoption of IRC by free software developers, the trend toward massive collaboration in virtual organizations, and the introduction of automation and abstraction mechanisms in the work process have all contributed to the continued popularity of the protocol within the free software community.

THE THREE PILLARS OF FUNCTIONAL AUTONOMY

In chapter 2, we argued that the capacity of a hacker community to resist recuperation attempts depends on three pillars of functional autonomy:

technical expertise, shared values, and historical memory. In retelling the story of IRC, we have focused on the role of technical expertise, as showcased by its frugal design. It allows hackers to use an antiquated protocol, overcoming its limitations and adapting it to their particular needs, taking advantage of its simplicity and flexibility. To balance the account, we want to end this chapter with a reflection on the importance of shared values and historical memory for sustaining the functional autonomy of IRC communities.

Shared values are closely tied to the protocol itself and make it meaningful as a piece of material culture. As one user put it, recalling a common trope on hacker forums, "IRC will never die, because it has culture" (ehmry, April 20, 2021). In the process of caring for their shared means of communication, users become emotionally invested in the valued attached to the medium. In a different context, Coleman emphasizes that the ethics and aesthetics of craftsmanship are closely intertwined in hacker cultures (2012). The utilitarian appearance of the text-only interface and the way in which it lends itself to custom postprocessing cannot be overestimated in this regard. Christopher Kelty discusses the interdependency between the media that hackers use and their community values under the heading of "recursiveness." Geek publics generate, in a recursive manner, the material conditions for their own continued existence through engagement with legal and technical practices and topics (Kelty 2005). Again, the engineering values of simplicity and flexibility manifest themselves in the text-only interface. Developers have consistently refused to incorporate multimedia features into the protocol throughout its long history, which preserved the advantages of the streaming protocol design, the main technical feature under consideration in this chapter. Ethical and aesthetic values come together in the asceticism of the terminal, promoting the economy of means and frugal engineering practices necessary to preserve the functional autonomy of communities in a technologically complex society.

Historical memory became an increasingly important element of IRC culture during the later period. The longevity of IRC today is a running joke in technology circles. IRC users may point out that new technologies provide little improvement over IRC, which is a tried-and-true medium that works for them. Even many nonusers of IRC recognize its use as a sign of principle and dedication to the old-school hacker ethos. IRC inspires a feeling of belonging, so much so that its contemporary competitors (listed later on in this chapter), however commercial, need to provide "bridges" to IRC if they purport to be viable products. Bridges allow IRC users to participate in chat rooms operated on other chat devices. Significantly, these bridges are necessary because IRC users would never cross them, only their messages would. As the web comic at the beginning of this chapter attests, the legend is that each technology company has at least one worker who is still using IRC, boycotting the transition to the new chat platform promoted by management.

MOMENTS OF CRITIQUE, RECUPERATION, AND COMMUNIZATION

Past attempts to recuperate IRC have failed in the sense that it has not generated any marketable product, while the infrastructure is continuously maintained by a thriving and self-governed hacker community. It amounts to a rare success story in our narrative about the dynamic of critique and recuperation. In part, this outcome is due to the frugality of the design, which in turn reflects a commitment among its administrators and operators to develop the technology in keeping with certain ethical and aesthetic values.

Although the IRC protocol itself has not been subsumed under a for-profit, corporate-controlled platform, we readily concede that the practice supports numerous peer production communities, which in turn are being valorized by capital at a higher level. Furthermore, the political culture spawned in connection with IRC extends to the troll tactics pioneered by Anonymous. These days, trolling has become an integral part of the election campaigns of far-right politicians, as well as a weapon in the arsenal of states waging information warfare. The ideological allegiances of the trolls are secondary to the fact that the service has been commodified and is up for sale on dark markets (Fish and Follis 2019). IRC contributes indirectly to the accumulation of capital, in much the same way as we previously argued in regard to hackerspaces. A wide range of institutional actors are seeking to leverage hackerspaces in order to promote industrial productivity, boost regional competitiveness, speculate in real estate development, and so on. An analogous argument can be made about free software development. As has been extensively documented elsewhere, computer firms rely on free software in order to cut their expenditure on basic infrastructure and programming labor.

Indeed, the wider relevance of IRC, as is the case with hackerspaces and free software, hinges on the fact that it provides a utility with which others want to engage, in order to further their diverging ends. Otherwise, IRC would have lacked societal relevance and been merely of archival interest, as some media scholars believe to be the case already. What counts is that the IRC protocol contributes positively to the struggles of hackers to preserve the functional autonomy of their other projects and communities. One reason that IRC fulfills this role is because it is possible to sustain it with such limited resources as can be procured from private donations or workplace embezzlement, occasionally with the tacit consent of management.

This exemplifies what we have elected to call the communization of the infrastructure. Communization is the antonym of privatization, and it describes the process whereby resources that are nominally managed under a corporate or government entity are mobilized to expand the functional autonomy of workers instead. It differs from the recycling of materials or the repurposing of tools in that it is inscribed within a contestation against the symbolic and material order of capital. Letting one's employer unwittingly pay for the upkeep of an IRC server is not as striking as the factory occupations of the Autonomia movement. Still, both cases qualify as communization in the sense that they demonstrate the repurposing of the means of production in a capitalist enterprise in order to serve a culture of worker autonomy.

CONCLUSION

The recuperation of critique is the historical logic of informational capitalism. Such a statement may sound overly cynical and a cause for despair. However, identifying a structural constraint—such as a historical logic—is not the same as declaring its invincibility. On the contrary, to have a conceptual understanding of the framing conditions of the present is a precondition for acting strategically and effectively.

The example of IRC demonstrates that the recuperative logic of informational capitalism can be, if not conclusively defeated, at least warded off for a very long time. The community of IRC administrators, operators, and users has withstood the recuperative logic of informational capitalism for several decades. Numerous, targeted recuperation attempts have been made over the years, without any spectacular success in bringing IRC under the aegis of innovation. The IRC protocol withstood the lure of the dot-com bubble that induced many other collaborative nonprofit development projects at the turn of the millennium to go commercial. After the burst of the bubble, the subscription model and the targeted advertising model emerged as the two enduring business models for delivering services over the internet, but the IRC networks remained free of charge and free of advertisements. In contrast to the other case studies discussed in this book, neither the technical invention (federated chat streaming protocol) nor the organizational idea (topic-centered pseudonymous conversations) have ended up under corporate ownership, and the community continues to thrive to this day.

IRC is so anachronistic in the present media landscape that one could conceive of it as a time machine, which takes its users back to the norms and values that prevailed on the internet in the early days. In the introduction, we quoted William Morris to make a point about how the struggles of past generations fade away, only to resurface in a different context and under different names. This observation captures the transition from analog radio to its digital successors in communication, including IRC, very well. Amateur radio enthusiasts and activists often mourn the shift to online radio. This sensation is reinforced by the utopian promises that were once inscribed in the practice and the sociability of CB radio. Its resilience to recuperation owed something to that, austere and restricted, design of analog radio. Present-day devotees of the IRC protocol express comparable scorn toward the adoption of corporate-controlled social media platforms.

The self-management of two-way communication in IRC networks has prevailed for an extended period of time. Although it can no longer rival the user base commanded by the commercial media outlets, as it could before, it has proven very resilient. This is a provisional and precarious victory. Nonetheless, it is a victory of strategic importance, since it allows hackers to self-organize the means of communication, upon which auxiliary development projects depend. It is a crucial stepping-stone for defending the functional autonomy of many other peer production communities. In ways that we cannot foretell at present, both as a communication tool and as a time machine experience, IRC will continue to inspire cycles of struggle to come.

7

CONCLUSION
THE STRUGGLE AHEAD

We opened this book with the claim that hacking is not primarily about technology. Rather, an idea of freedom propels the urge for change. Hacking holds out the possibility that such freedom can be achieved by repurposing tools and circumventing constraints of different kinds. Tied to this practice is the acclaimed outsider position of the hacker. Popular culture is intoxicated by tropes about the hacker as an outcast. Self-representations of hacker culture are awash with the same outlaw imagery, a good example being the Jolly Roger flag that decorates hackerspaces and makerspaces and their associated websites. A case in point is the emblem of the Chaos Computer Club, designed by Way Holland in 1981, which distorts the horn-and-lightning logo of the Deutsche Bundespost (German Federal Postal Service) to make it look like a pirate's cranium.

The catch is that the symbolic outsider position of the hacker is increasingly at odds with the structural position of hacking within present-day informational capitalism. The plausibility of the hacker's claim to be an outsider rests on the position of the hacker standing outside of contractual employment relations and the associated, professional identities. In the early days of the computer underground, regulated and lifelong employment was the norm in society. Those who opted out from this legally secure arrangement were true outliers. But over the last thirty years or so, an ever larger segment of the workforce has had to make a living under the same precarious conditions as the hacker. In the so-called "gig economy," everyone is an outsider.

"Open" is the sibling word to "outside." Openness is the means by which the outsider becomes included in capital's accumulation regime. As capitalism restructures itself around open innovation processes, the hacker becomes emblematic for how value is produced and captured everywhere in this economic system. The disruptive hack has always already been anticipated in this open innovation model. Whole academic fields are developing and refining the methods whereby firms can "harness the hacker" in ever more cost-efficient and risk-averse ways (Tapscott and Williams 2006; von Hippel 2016; Flowers 2008 are representative examples). Consequently, the promise that the repurposing of technology will unsettle constituted power and incumbent interests serves as a honeytrap for idealistically minded engineers. Their longing for freedom is absorbed and fed back into the maelstrom of Schumpeterian creative destruction.

As the methods for capturing value from open innovation processes are increasingly anticipated and refined by both firms and academics, the meaning of hacking is also undergoing a transformation. From this follows the need to reassess the way in which this subject matter is studied. Forty years ago, the original ethnographic works about hackers, notably Steven Levy's classic *Hackers: Heroes of the Computer Revolution* (1984), offered peepholes into an otherworldly and self-enclosed cosmos. When we peep through that same hole today, what we see is our own future selves, at work. The future, unevenly distributed as we know it to be, is disclosed to us in the utopias of hackers. The catch is, however, that those visions tend to be realized in an inverted and nightmarish form. This idea has founded the methodological assumption behind the studies in this book. Hacker culture can be interrogated as an early warning system of the structural reforms of the labor market that are yet to be unleashed on the rest of us.

Of course, having such warnings would be pointless unless it was also possible to identify points of intervention. Ultimately, the purpose of theorizing hacking, or, indeed, theorizing anything at all, is to guide collective action. The anticipation of the disruptive hack must itself be anticipated, for the hack to result in something truly disruptive. To anticipate the anticipation requires theory. For instance, the theoretical concept of recuperation alerts us to the traps of the purported outsider position ascribed to the hacker in popular culture. In keeping with the tradition of Hegelian and immanent critique that underpins the present inquiry, we track down the points of possible intervention to cracks and fissures within the contradictory social totality.

To substantiate this rather lofty claim, recall the foundational move of the free software movement, the creation of the General Public License (GPL). With this legal hack, copyright law was turned against itself to protect the information commons from exclusive proprietary claims. The prerogatives invested by state and law in the individual author, with the expectation that authors' rights would in a short time be alienated and put up for sale, were thus redirected toward collective authorship of source code. That being said, if one puts too much trust in the protection of the law from itself, then one risks succumbing to legal formalism. The enforcement of the GPL relies on it being backed up by community norms. Those norms need to be in place in order to ensure that individual members and would-be entrepreneurs fall back into line at critical junctures during the course of a hacker project—that is to say, at times when business opportunities loom large or legal deterrents weigh in on collective decision-making processes. The principal importance of the GPL is as a rallying point for collective action. From this, it follows that it is not in the absence of society, but quite the opposite, in the plethora of the social bonds and the commitment to common goals, that the conditions of freedom may flourish under the constraints of class antagonism, exploitation, and commodification.

In invoking the functional autonomy of hacker communities, we are indebted to how this concept has been deployed in the literature of labor process theory. The concept of "autonomy" is decisive in the distinction, proposed by Karl Marx in the "Results of the Direct Production Process," between the formal and the real subsumption of labor (1994). A total domination by capital over the production process (i.e., the real subsumption of labor under capital) is checked by the functional autonomy and effective control that workers exercise collectively over their workplaces, in part by possessing knowledge about how to put their tools to productive use (Aronowitz 1978). The historical parallel between struggles over tools and skills on the shop floor and such struggles unfolding in hacker culture at present is justified by the observation made above: having been put to work by the open innovation model, hackers are thenceforth formally subsumed under capital. The recursiveness attributed to hacker politics, by which is meant their orientation toward safeguarding the technical and legal preconditions for their own continued, collective existence *qua* hackers, resonates with this analysis. Hackers are called to action in response to an incessant pressure to subsume hacking under processes of capitalist accumulation.

The political orientation of hackers is toward the preservation of their own autonomy. That being said, the outcome of those struggles often has consequences for many communities and sectors beyond hacker culture. On those struggles depends the extent to which hacker projects can nurture critical engineering practices and processes of communization. Conversely, the alternative pathways in the development of technology are narrowed down when hackers fail to resist recuperation attempts.

An advantage of adopting this interpretative framework is that it steers a middle course in the dispute between one camp in academia that debunks the technological determinism of hackers and their many privileges and another camp that boosts the promises and hopes (and hype) surrounding information technologies and so-called "making." We do not side with one or the other of these camps. Our proposal is instead that the meaning of a hacker practice or application depends on the outcome of struggles against recuperation. Furthermore, this question cannot be decided once and for all, since both recuperation processes and the resistance to them are always only inconclusively settled. Each advance or defeat lays the groundwork for the next cycle of struggle. With a nod to Marx's eleventh thesis on Feuerbach, we intend this interpretative framework to not only describe hacker culture, but also lean in on how hackers narrate their own past and orient themselves toward the future.

This is effectively what scholars in the field of innovation studies do already. They intervene by advising firms on how to fine-tune the open innovation model and harness the hacker ever more cost-effectively. We hope that hackers will find ideas in this book for how they can counter such approaches. Theorizing recuperation makes a difference because recuperation often works by stealth, behind the backs of the actors, as it were. An attempt at recuperation can easily be detected in those instances when it is carried out as a discrete enclosure of the information commons and in open violation of community norms. It is harder to challenge the relentless and gradual pressure on the community's norms and goals, whereby the framing conditions of individual hacker projects become aligned with the requirements of the open innovation model. As the hacker community is made over, the development processes nurtured in this milieu will come to gravitate toward market demand, mass production constraints, and various kinds of legal deterrents.

The recuperative logic of informational capitalism is hard to glimpse in the local setting because the actual act of enclosure of the information commons might still be pending. For instance, the ever-growing capacity to aggregate and triangulate user data enables firms to assert control over and extract revenue from information in ways that were impossible to foresee twenty years ago. The continued expansion of such capacities and opportunities is anticipated by firms and feeds back into their present-day market strategies. Struggles over recuperation that unfold on this strategic and anticipatory level are located within the second time horizon in our classificatory scheme. It is chiefly when contesting the recuperation that takes place at this level of temporality that the interpretative framework proposed here could make a difference.

The example of how big data and AI allow preexisting datasets to be mined in ways that were inconceivable at the time when the data was first collected concretizes a more general tenet about recuperation. Namely, that recuperation works by surprise. This was William Morris's core insight as conveyed in the quote cited at the beginning of this book. In his account of historical struggles, Morris laid stress on naming practices. What is meant by a name is in a state of flux. The same goes for the (theoretical) concepts and categories by which we try to identify and put a name to "recuperation." Those categories are caught up in the same whirlwind of capitalist restructuring as that affecting the things to which they refer. This point bears stressing. When the meaning of a concept is taken to be fixed and a matter of course, then this is an invitation to unpleasant surprises.

Differently put, recuperation cannot be exhaustively summarized in a bullet point list. Hence, the discussion of theory needs to be complemented with the study of historical cases and experiences derived from past struggles. In saying this, we are freely reinterpreting Immanuel Kant's famous notes on "reflective judgment." In the introduction to his third critique, and as a justification for the aesthetic theory that followed immediately after, he conceded the limitations of predicatively structured reasoning. The exploration of the world could not do without the cognitive subject's ability to form higher-order analogies drawing on contextualized experiences. It was thus that the subject made the jump from knowledge of the past to judicious, future-oriented action. In the context of our present inquiry into hacker culture, we reinterpret Kant to say that it is by cultivating an aesthetic judgment that hackers may learn to "sniff out" ongoing recuperation attempts. Ultimately, we have written this book with the aspiration of making a contribution to this exercise. It follows that the interpretative framework proposed here must be put to the test, in part through a discussion of historical case studies, in part through praxis (Kovel 2008).

LESSONS FROM FOUR HISTORICAL CASE STUDIES

The four case studies presented in this book demonstrate different possible outcomes in the struggle over the recuperation of hacker practices. In addition, the four cases illustrate the different time horizons within which struggles over recuperation unfold. Overall, we historicize the case study genre through shifting the analytical attention away from snapshots of the present toward tendential developments in hacker culture unfolding over extended temporalities.

The first case study, concerning the Ronja project, is centered on a schism in the hacker community that was triggered by attempts to commercialize free space optics (FSO) technology. Although there were many proprietary spin-offs from the invention, the most ambitious and enduring enterprise went by the name "Crusader." Ronja and Crusader are interesting to compare because both development projects were informed by articulated political visions. Hence, they can be treated as representatives of divergent ideas about how to do politics through engineering, and what role to assign to the market in this scheme. For the inventor of Crusader—as for a large segment of the Czech wireless network community at the time—the monopolistic practices in the telecommunications sector were the principal targets. The objective was to challenge the monopoly by equipping wireless network communities and for-profit internet service providers with Crusader FSO links. The communication network as a whole would become more resilient to state censorship and surveillance through the diversification of providers. This blend of Schumpeterian entrepreneur-ialism with political activism has gained much traction over the years, not only in the hacker movement, but also in technology- and product-oriented social movements in general (Hess 2005).

In contrast, the inventor of Ronja was convinced that devising an alternative business model would not be conducive to his political vision. His ideal of a user-controlled technology was antithetical to the very idea of the market. If he took the customary route of financing Ronja develop-ment through the market and the patent system, the technology would end up perpetuating the ills of commercial development that it was meant to combat in the first place. He sought to make his development work economically sustainable by asking for donations from the many users of the product. The donation model provides a touchstone for reflecting upon how technology development can be scaled up without relying on the conventional, market- and industry-based means for doing so. A hand-ful of updates to Ronja were successfully financed through donations. The model hinged on the reputation of the inventor within the Czech wireless network community. Donations stopped coming in as the community dis-integrated. This outcome was in part due to shifts in the policy landscape, which led to communication networks becoming widely available as a commercial service. In part, the user-controlled technology failed to keep up with commercial developments due to escalating, internal strife among the hackers over conflicting proprietary claims. In relation to our interpre-tative framework, we could say that the recuperation attempt of Ronja resulted in a draw. The hacker community imploded, and the develop-ment of the technology stalled, but neither was any commercially viable innovation derived from the proprietary forks.

In the second case study, we tell the story of the open-source desktop 3D printer called RepRap. Initially, the development process was self-organized by a community of hardware hackers and engineering students. The attempt by one start-up company, MakerBot Industries, to enclose the pool of common labor behind proprietary claims was a watershed moment. It qualifies as a textbook example of an overt, hostile recuperation attempt. In spite of the fuss that ensued in the extended open hardware move-ment, the violation of the free license was allowed to pass in the RepRap community. This indicates to us the importance of community norms for the protection of the information commons. We concede that the permissive attitude toward commercialization might have prevented the RepRap community from being torn into warring factions, as happened with Ronja in the previous case study. Nevertheless, soon afterward, the development of the open-source 3D printer came to a near standstill. Hackers in the core team silently dropped out of the project or channeled their efforts into private ventures. A couple of years down the road, most of the start-up firms had gone out of business, while others had been acquired by multinational corporations. This outcome contrasts sharply with the goal that was stated at the outset of the project: to devise a self-reproducing, general-purpose manufacturing unit that would render market exchanges superfluous and create wealth without money.

The flourishing consumer market in desktop 3D printers is a telltale sign that a successful recuperation attempt took place. That being said, the product innovation itself is rather insignificant compared to the organizational concept that lay at the heart of the RepRap project. The key idea behind the self-replicating universal constructor was that its manufacturing capacities would be recursively deployed in the production of more such units. MakerBot Industries put out feelers in this direction by enrolling the machine park of its former customers in the production of parts for new machines. The experiment was prematurely ended due to the lack of a means for asserting worker discipline (quality control) in the dispersed production network. Shortly afterward, the possibility of reducing production costs through open manufacturing was foreclosed by the company's decision to revert to a conventional business model based on closed designs and intellectual property rights. Due to the contradictions of the open accumulation regime, the industry failed, at least for the time being, to incorporate the hacker community's organizational inventions.

The third case study unfolds within the second time horizon and relates to a movement of numerous individual hacker projects. This story charts a series of transmutations of shared machine shops—the generic name we assign to physical spaces where manufacturing tools are made available to the public—that have unfolded over the last two decades. The successive order began with hacklabs, later to be replaced by hackerspaces, which soon afterward were rebranded as makerspaces; makerspaces inspired Tech Shops, and Tech Shops gave way to start-up incubators. The series

ends in the accelerators. All of these settings serve to facilitate the circulation of tools and technical know-how outside the confines of professional
identities and hierarchies. This commonality underlines the single trait
that separates them—namely, the extent to which an autonomous political culture and critical engineering practice may flourish within a certain
setting.

Hacklabs are located in occupied social centers. As such, they are both
ideologically and spatially integrated into anarchist or autonomist social
movements. One common task of hacklab participants is to support other
political activists at the social center through maintaining and developing
digital infrastructures. The heyday of hacklabs was during the early 2000s,
with a geographical concentration around southern Europe. In contrast,
hackerspaces are nonpartisan clubhouses for the cultivation of technological creativity. Typically, they are located in rented spaces financed through
membership fees. A proportion of their members identify themselves as
activists, who pursue various sorts of civil society activities. Historically,
the upsurge of hackerspaces in many European and US cities can be dated
to the 2010s.

Our central claim is that these different genres of shared machine shops
evolved in tandem with a gradual buildup of recuperation, culminating in
the start-up accelerator. This claim can most succinctly be demonstrated
with reference to the different roles that these spaces have played in urban
planning and real estate development. A tactic used by antisquatting agencies is to allow a few tenants to rent empty spaces below market price in
what would otherwise have been empty buildings. Among such tenants
are sometimes found hackerspaces. On the surface, much will look the
same in a shared machine shop that is housed in a squatted building and
one that rents the space from an antisquatting agency. And yet, the political significance of these two sites is diametrically opposed. One lesson
that we draw from this case study is that making tools and skills accessible
is insufficient on its own to nurture critical engineering practices. The
other two "pillars of autonomy" must also be in place: the maintenance of
shared values and the reproduction of a historical memory of past events
and struggles.

The fourth case study also runs its course within the second time horizon,
involving a landscape of sedimented infrastructures and communication

protocols that make up the framing conditions for many contemporary, individual hacker projects. In this case, however, the hackers have successfully resisted attempts to incorporate their technology under a regular business arrangement. The case study focuses on Internet Relay Chat (IRC), a simple but flexible protocol for real-time, written conversations. It was first implemented in 1988, one year before the World Wide Web was launched. It served various functions that were later spun off and performed on dedicated, corporate-owned platforms, such as online dating, staying in contact with friends, and file sharing.

As the population of the internet grew and market consolidation set it, IRC faded from public view. It continues to be used for everyday backstage communication by free software development communities (Coleman 2012) and among hackerspace members (Maxigas 2015), Wikipedia editors (Broughton 2008), and Anonymous hacktivists (Dagdelen 2012). Therefore, the question arises: Why do these groups of users—hailed as disruptive innovators and early adopters—stick with a chat technology that has become a museum piece, despite its technical limitations within the current media landscape? It suggests to us that the regular criteria for technology adoption, such as user convenience, functionality, and efficacy, all of which are heavily skewed in favor of network effects and the size of capital, can be overridden by ethical and aesthetic judgments in settings where the three pillars of autonomy are all in place.

Overall, we note that the first and second case studies follow the life cycles of individual projects, unfolding within the first time horizon, whereas the third and fourth case studies are located within the second time horizon of an evolving movement of hacker projects. The struggle over recuperation in the Ronja project ended in a stalemate in the sense that no hacker project has built on its results, but neither did the proprietary forks lead to any commercially viable innovations. The second case study of the RepRap 3D printer offers a textbook example of a hacker project that succumbs to a hostile recuperation attempt and seeds product innovations for capital as a result. The conclusion is much the same in the third case study of shared machine shops, with the difference being that recuperation did not culminate in a tangible, commodified product, but in a shift within hacker culture, creating favorable conditions for future recuperation attempts. The fourth case study of IRC demonstrates that hackers occasionally succeed in resisting the historical logic of recuperation over an extended period of time. This underlines the difference that collective action can make, even though it takes place under the structural constraints of informational capitalism.

Recuperation, being an evasive phenomenon, is often only detectable with the benefit of hindsight. That being said, the purpose of studying history is to draw lessons from it to guide actions in the present, with an eye on the future. Hence, we close this book by putting our theoretical and historical arguments to the test by attending to events that are unfolding even as we type these sentences.

THE NEXT FRONTIER: HACKING PHARMA

The outbreak of the COVID-19 pandemic in 2020 has brought some vulnerabilities in the capitalist economy to public attention. Confidence in global value chains was shaken when nation-states began to seize shipments of medical supplies, such as, for instance, face masks, destined for other countries. Another eyeopener has been the outspoken negligence of some heads of state, notably in the United States, Brazil, and Belarus, to act on the best available epidemiological knowledge in order to limit the spread of this infectious disease in their populations. From the highest quarters of political office, the message was sent to citizens that "you are on your own now." It is a shocking message, if only because it was so indiscriminately broadcast. Even that segment of the citizenry that was formerly enfranchised by a healthcare system is henceforth exposed to the kinds of biomedical risks that unfranchised groups have grown accustomed to over the years.

What lessons the public will draw from the policy failures related to COVID-19 remain to be seen. Right at the outset of the pandemic, hackers were summoned to restate their old case for local self-reliance and decentralized supply and production chains. One DIY response to the pandemic has been the flooding of 3D design repositories with printable face shields and similar kinds of medical trivia. The expansion of hacker culture in the direction of open pharma will be hastened by this global health crisis, although this trend has already been waiting in the wings for a number of years.

Hacking pharmaceuticals is the logical next step for the biohacker movement, which branched off from hacker culture a decade ago. Biohacking emerged in tandem with the dissemination of cheap and user-friendly wetware laboratory instruments (Delfanti 2013). The punk aesthetic and underdog rhetoric championed by spokespersons of the incipient biohacker movement contrasted markedly with their eagerness to comply with predominant epistemological hierarchies and regulatory protocols. In the words of one unsympathetic observer, DIY bio quickly became the "backyard of the biotech industry" (Ikemoto 2017). A catalyzing factor was the alarm in the media over public security. Such anxieties tend to be aroused by hacking in general but were strongly reinforced by the frightening scenario of bioterrorism. Out of concern for its public image, biohacking community leaders teamed up with the FBI at the first opportunity. This alliance bestowed the biohackers with credibility and made them into eligible recipients of private investment and public funding. Tocchetti and Aguiton (2015) conclude their study with an ironic twist. Due to police involvement in the name of public security, DIY biolabs speedily evolved into becoming complicit in the institutionally sanctioned forms of hazardous and irresponsible innovation that goes on in the biotech industry.

When some biohackers begin to explore DIY medicine production, the question looms large as to whether this new field of investigation will succumb just as quickly to the mandates of state and industry. Will the medical branch of the hacker movement be turned into the pharmaceutical industry's backyard? If loathing of Big Pharma offered any guarantees, then the answer would be a resounding "no." The staggering malpractices in the sector, which have been extensively documented by scholars over many years (Abraham 2008; Mirowski 2011, chap. 5; Rajan 2017), are explicitly targeted by high-profile pharma-hacker projects. For instance, the Open Insulin Project was triggered by the extortionate pricing schemes for this lifesaving compound (Gallegos et al. 2018), and the Open Source Malaria project seeks to fill a void in medical research due to the lack of commercial interest in curing tropical diseases (Arza and Sebastian 2018). The Four Thieves Vinegar Collective entered the fray by reverse-engineering and disclosing to the public a lifesaving cure for parasitic infections called pyrimethamine, shortly after its price had been raised from $13.50 to $750 per dose by the holder of the intellectual property rights. The narrative of David versus Goliath is part of the standard repertoire of hacker culture, and, as the above examples attest, these roles were frequently enacted by the nascent biohacker movement too.

The spokesperson for the Four Thieves Vinegar Collective gave the following explanation as to why the biohacking movement had caved in as soon as it came under pressure. The movement around biohacking was summoned almost from one day to the next, he recalls, attracting members with only a superficial allegiance to the hacker ethic. This contrasts unfavorably with the hacking of computers in the early days, which only caught the public's attention after a prolonged period of subcultural gestation. Furthermore, the first wave of software hacking benefited from having organic intellectuals among their ranks, who could articulate goals and clarify the rationales of the hacker movement (Mixael Laufer, September 4, 2020).

The above remarks resonate with our own observations of earlier controversies within hacker communities. Whereas FBI agents were welcomed to the biohacker movement without it causing much fuss, the decision by the organizers of a Dutch hacker camp in 2013 to let the police set up a tent at the site was met with fierce resistance (Maxigas 2014). The contestations originated in a core of old-school hackers whose involvement in the scene went back to the first hacklabs, established in the 1980s with close ties to the squatting movement.

The lesson about the importance of historical memory does not inspire confidence in the future trajectory of pharma-hacking. That being said, Four Thieves Vinegar Collective is cooperating with harm reduction activists and reproductive health activists on projects where their goals overlap. The former has a stake in, to give but one example, the diffusion of methods for the DIY manufacturing of an antidote drug against opioid overdosing (naloxone). Likewise, the latter welcomes information being disclosed to the public about how to manufacture abortion medication at home (such as, for instance, mifepristone). This suggests that the expansion of hacking to the field of medicine will have the secondary consequence that hacker culture will be exposed to new influences from established and often militant social movements. Arguably, this is analogous to the exchange of ideas and political values that once stemmed from the symbiotic relationship between hacklabs and the squatter movement.

Based on the discussions above, we believe that the outcome of the latest cycle of hacker struggles will depend on them successfully "anticipating the anticipation." It is foreseeable that the pharmaceutical industry, which is well-versed in capturing patient groups and using them in its marketing and lobbying strategies (Mulinari et al. 2020), will try to do the same with hackers. Innovation scholars are standing by, ready to give pharmaceutical companies advice on how to extract ideas and mine data from users conducting nonauthorized experiments with the firms' products. To ensure that the latest dream of freedom is not realized in a nightmarish form, pharma-hackers must prepare themselves for the encroachment that is coming, in order to resist being subsumed under an open innovation model.

NOTES

CHAPTER 5

1. A full list of hackmeetings is reproduced in Maxigas (2015, 125).

2. Richard Stallman attended the Italian hackmeetings in 2002 (Bologna) and 2011 (Florence). He also gave a speech in the Casa Invisible occupied social center in Malaga, Spain (Lee 2008). Most recently, hackers phoned him for a conversation on community wireless networks during the Iberian hackmeeting (2014, Marinaleda, Spain) at which one author was present.

3. One of the authors lived in an occupied social center called "Non-Commercial House" on Commercial Road, London, mainly dedicated to serving as a freeshop, strategically located in a shop at the intersection of the financial district, an immigrant neighborhood, and a gentrified hipster area (The City, Whitecross, and Shoreditch, respectively).

4. This section draws on fieldwork with participants at the Print Hacklab in the Les Tanneries occupied social center in Dijon from July to September 2006 and retrospective interviews with the hacklab participants (Lunar, May 23, 2013; Darkveggy March 22, 2014; Ricola, February 1, 2014). These experiences are recounted in more detail in Mauvaise Troupe (2014).

5. Escamot Espiral struck on October 13, 2007, in the CSO Astra in Gernika-Lumo during the Iberian hackmeeting and on January 29, 2008, in Espai Obert, the editorial headquarters of the Contrainfos political bulletin.

6. This section draws on interviews with two Escamot Espiral members (Dulzet and Joseba, August 16). The actions are documented on the archived website of the Swirl Commando (Escamot Espiral, 2008).

7. This section draws on fieldwork at the Forte Prenestino occupied social center in Rome in June 2006 during the Transmission video activist meeting and a set of

interviews with members of the Autistici/Inventati collective in Dublin (June 2013). These experiences are recounted in more detail in Beritelli (2017).

8. The Chaos Communication Camp takes place every four years. There are also similar camps organized in the Netherlands in a complementary four-year cycle. Both strings of events complement the annual Chaos Communication Congress, as their (even more) community-oriented counterpart. These events together constitute the major rhythm of what we call the Northern Circuit of Hacking.

9. "Die Schleuse und die Mainhall stehen allen Lebensformen zur Verfügung" (translation by the authors).

10. These have been extensively surveyed and described in Maxigas (2015, chap. 10).

11. The artefact is documented on the wiki of the hackerspace under the following address: https://wiki.techinc.nl/PCB_Techinc_Logo

12. This section draws on long-term fieldwork in London and Budapest while one author lived in these cities and frequented these hackerspaces. These field experiences have not been documented in previous works in any detail.

13. Programming paradigms identify rival philosophies in language design, writing, and organizing source and are different from the scientific paradigms famously proposed by Kuhn (1962).

CHAPTER 6

1. Latzko-Toth (2010) writes about "metaphors of synchrony," including radio, telephone, and conferences. The radio metaphor was already popular before the internet. "Channels" were used on the PLATO system; CompuServe's 1980 CB Simulator chat service started with forty channels like contemporary CB radio networks, while the name of BITNET Relay referred to radio relay towers. On the internet, a limited but popular chat system was called ICB, which stood for Internet Citizen's Band. The latter emerged in 1989, more or less in parallel with IRC, the subject of this chapter.

2. Very high frequency (VHF) bands have considerably more channels than CB radio, and VHF is used, for instance, in maritime communications.

3. Grier and Campbell (2000) explain that even BITNET Relay relied on the store-and-server model. Links between nodes in the BITNET network were unstable, so BITNET Relay kept them until the link became operational again. However, it was similar to the "relay" in IRC in the sense that the purpose was to distribute messages between different servers connecting a geographically dispersed user base, in the fashion of relay towers in amateur radio. IRC works via the internet, so it can delegate the problem of reliability to the underlying Transmission Control Protocol (TCP).

4. Splits in the social life and social history of IRC are in many ways equivalent to the more widely theorized "forks" in free and open-source software development (Raymond 1999; Weber 2004; Coleman 2012), which makes sense, given that software development was a familiar practice to many users of IRC during the early internet era.

5. It is crucial to remember that, before Web 2.0, interactivity on the World Wide Web was rather limited, so that chat felt even more real time than it does today.

6. These assertions are made with reference to a website that provides statistics on IRC use since August 1998: http://irc.netsplit.de/networks/top10.php.

7. Due to these discussions, such software is often treated under the general rubric of Free, Libre, Open Source Software (FLOSS). We address some of these controversies in the introduction.

REFERENCES

Abelson, Harold, and Gerald Jay Sussman. 1996. *Structure and Interpretation of Computer Programs*. Second ed. Electrical Engineering and Computer Science. Cambridge, MA: MIT Press.

Abraham, John. 2008. "Sociology of Pharmaceuticals Development and Regulation: A Realist Empirical Research Programme." *Sociology of Illness and Health* 30 (6): 869–885. https://doi.org/10.1111/j.1467-9566.2008.01101.x.

Ackermann, John R. 2009. "Toward Open Source Hardware." *University of Dayton Law Review* 34 (2): 183–223. https://www.tapr.org/Ackermann_Open_Source_Hardware_Article_2009.pdf.

Aglietta, Michael. 1976. *A Theory of Capitalist Regulation: The US Experience*. London: Verso.

ana. 2004. "Hacklabs: From Digital to Analog." Blog entry, translated from the Suburbia Telemacktical MediaZine. https://network23.org/ana/hacklabs-from-digital-to-analog/.

Anarchopedia contributors. 2006. "LOA Hacklab." Encyclopedia entry. "storia." 2010. http://eng.anarchopedia.org/index.php?title=LOA_Hacklab&oldid=19251.

Andersson, Jonas. 2011. "The Origins and Impacts of the Swedish Filesharing Movement: A Case Study." *Journal of Peer Production* 1 (1): 1–18. https://www.diva-portal.org/smash/get/diva2:698889/FULLTEXT01.pdf.

Antonio, Robert J. 1981. "Immanent Critique as the Core of Critical Theory: Its Origins and Developments in Hegel, Marx and Contemporary Thought." *The British Journal of Sociology* 32 (3): 330–345.

Aronowitz, Stanley. 1978. "Marx, Braverman and the Logic of Capital." *Critical Sociology* 8 (2–3): 126–146. https://doi.org/10.1177/089692057800800214.

Arza, Valeria, and Sol Sebastian. 2018. "Open Source Pharma and Its Developmental Potential." *Liinc Em Revista* 14 (1): 47–64. https://doi.org/10.18617/liinc.v14i1.4144.

Atton, Chris. 2004. *An Alternative Internet: Radical Media, Politics and Creativity.* Edinburgh: Edinburgh University Press.

Auray, Nicolas, and Danielle Kaminsky. 2007. "The Professionalisation Paths of Hackers in IT Security: The Sociology of a Divided Identity." *Annales Des Télécommunications* 62 (11–12): 1312–1326. https://doi.org/10.1007/BF03253320.

Barberi, David. 1999. "Communication Archive." FTP directory on ibiblio.org. http://ibiblio.org/pub/academic/communications/.

Barbrook, Richard. 2006. *The Class of the New.* London: OpenMute. http://web.archive .org/web/20111117001356/http://www.theclassofthenew.net/the_class_of_the_new .pdf.

Barbrook, Richard. 2007. *Imaginary Futures: From Thinking Machines to the Global Village.* London: Pluto Press.

Barbrook, Richard. 2015. "Cyber-Communism: How the Americans are Superseding Capitalism in Cyberspace." In *The Internet Revolution: From Dot-Com Capitalism to Cybernetic Communism,* Vol. 10:28–51. Network Notebooks. Amsterdam: Institute for Network Cultures.

Barbrook, Richard, and Andy Cameron. 1996. "The Californian Ideology." *Science as Culture* 26: 44–72. http://www.imaginaryfutures.net/2007/04/17/the-californian -ideology-2.

Bardzell, Shaowen, Lilly Nguyen, and Sophie Toupin. 2016. "Special Issue: Feminism and (Un)hacking." *Journal of Peer Production,* no. 8. http://peerproduction.net/issues /issue-8-feminism-and-unhacking/feminist-hackingmaking-exploring-new-gender -horizons-of-possibility/.

Barron, Anne. 2013. "Free Software Production as Critical Social Practice." *Economy and Society* 42 (4): 597–625. https://doi.org/10.1080/03085147.2013.791510.

Bazzichelli, Tatiana. 2009. *Networking: The Net as Artwork.* Aarhus, Denmark: Digital Aesthetics Research Center, Aarhus University.

Benkler, Yochai. 2006. *The Wealth of Networks: How Social Production Transforms Markets and Freedom.* New Haven, CT: Yale University Press.

Beritelli, Laura, ed. 2017. *+Kaos: Ten Years Hacking and Media Activism.* Amsterdam: Institute for Network Cultures. https://networkcultures.org/blog/publication/kaos -ten-years-of-hacking-and-media-activism/.

Bey, Hakim. 1991. *T.A.Z.: The Temporary Autonomous Zone, Ontological Anarchy, Poetic Terrorism.* New York: Autonomedia.

Bisanz, Charles F. 1977. "The Anatomy of a Mass Public Protest Action: A Shutdown by Independent Truck Drivers." *Human Organization* 36 (1): 62–66. https://www .jstor.org/stable/44125241.

Blussé, Julie. 2013. "Twitter's Activist Roots: How Twitter's Past Shapes Its Use as a Protest Tool." *Radio Netherlands Worldwide*, November 15, 2013. https://www.pressenza .com/2013/11/twitters-activist-roots-twitters-past-shapes-use-protest-tool/.

Böhm, Steffen, and Chris Land. 2012. "The New 'Hidden Abode': Reflections on Value and Labour in the New Economy." *Sociological Review* 60 (2): 217–240. https:// doi.org/10.1111/j.1467-954X.2012.02071.x.

Bologna, Sergio, and Francesco Ciafaloni. 1965. "I Tecnici Come Produttori e Come Prodotto [The Technicians as Producers and Product]." *Quaderni Piacentini* 37 (March): 52–71. https://bibliotecaginobianco.it/flip/QPC/08/3700/#54.

Boltanski, Luc, and Eve Chiapello. 2005. *The New Spirit of Capitalism*. New York: Verso.

Bone, Jonathan, Juanita Gonzalez-Uribe, Christopher Haley, and Henry Lahr. 2019. "The Impact of Business Accelerators and Incubators in the UK." *BEIS Research Paper Number 2019/009*. https://assets.publishing.service.gov.uk/government/uploads/system /uploads/attachment_data/file/839755/The_impact_of_business_accelerators_and _incubators_in_the_UK.pdf.

Borland, John. 2007. "'Hacker Space' Movement Sought for US." *WIRED Magazine*, August 11, 2007. http://www.wired.com/2007/08/us-hackers-moun/.

Bosqué, Camille. 2015. "What Are You Printing? Ambivalent Emancipation by 3D Printing." *Rapid Prototyping Journal* 21 (5): 572–581. https://doi.org/10.1108/RPJ-09 -2014-0128.

Bratton, Benjamin H. 2015. "Platforms." Chap. 9 in *The Stack: On Software and Sovereignty*, 41–45. Boston: MIT Press.

Braverman, Harry. 1974. *Labor and Monopoly Capital*. New York: Monthly Review.

Bre and Astera, eds. 2008. "Hackerspaces: The Beginning." https://blog.hackerspaces .org/2011/08/31/hackerspaces-the-beginning-the-book/.

Brecht, Bertolt. 1964. "The Radio as an Apparatus of Communication." In *Brecht on Theatre: The Development of an Aesthetic by Bertolt Brecht*, edited by John Willett, 51–53. Frankfurt am Main: Suhrkamp Verlag.

Brodkin, Jon. 2016. "The WRT54GL: A 54Mbps Router from 2005 Still Makes Millions for Linksys—Open Source Firmware, Reliability Makes an Old Product Attractive to New Buyers." *Ars Technica*, July 1, 2016. https://arstechnica.com/information -technology/2016/07/the-wrt54gl-a-54mbps-router-from-2005-still-makes-millions -for-linksys/.

Burawoy, Michael. 1979. *Manufacturing Consent: Changes in the Labor Process under Monopoly Capitalism*. Chicago: University of Chicago Press.

Carpentier, Nico. 2008. "The Belly of the City: Alternative Communicative Networks." *The International Communication Gazette* 70 (3–4): 237–255. https://doi.org/10.1177 /1748048508089950.

Castells, Manuel. 1996. *The Rise of the Network Society, the Information Age: Economy, Society and Culture.* Vol. 1. Oxford: Blackwell.

Cavalcanti, Gui. 2014. "Is It a Hackerspace, Makerspace, TechShop, or FabLab?" *Make: Magazine,* May 22, 2014. http://makezine.com/2013/05/22/the-difference -between-hackerspaces-makerspaces-techshops-and-fablabs/.

c-base wiki contributors. 2019. "CIMP: C-Base Initial Member Package." Wiki page. https://wiki.c-base.org/dokuwiki/cimp.

Chesbrough, Henry William. 2003. *Open Innovation: The New Imperative for Creating and Profiting from Technology.* Watertown, MA: Harvard Business Review Press.

Chester, Andrea, and Carina B. Paine. 2007. "Impression Management and Identity Online." In *Oxford Handbook of Internet Psychology,* edited by Adam Joinson, Katelyn McKenna, Tom Postmes, and Ulf-Dietrich Reips, 223–236. Oxford: Oxford University Press.

Chopra, Deepak, and Scott D. Dexter. 2008. *Decoding Liberation: The Promise of Free and Open Source Software.* London: Routledge.

Cleaver, Harry. 2016. "Circuits of Struggle?" *The Political Economy of Communication* 4 (1): 3–34. http://polecom.com/index.php/polecom/article/view/62.

Cleaver, Harry. 2017. *Rupturing the Dialectic: The Struggle Against Work, Money, and Financialization.* Chico, CA: AK Press.

Cockburn, Cynthia. 1985. *Machinery of Dominance: Women, Men and Technical Know-How.* London: Pluto Press.

Coleman, Gabriella. 2004. "The Political Agnosticism of Free and Open Source Software and the Inadvertent Politics of Contrast." *Anthropology Quarterly* 77 (3): 507–519.

Coleman, Gabriella. 2010. "The Hacker Conference: A Ritual Condensation and Celebration of a Lifeworld." *Anthropological Quarterly* 83 (1): 47–72. http://muse.jhu.edu /journals/anq/summary/v083/83.1.coleman.html.

Coleman, Gabriella. 2012. *Coding Freedom: The Ethics and Aesthetics of Hacking.* Princeton, NJ: Princeton University Press.

Coleman, Gabriella. 2014. *Hacker, Hoaxer, Whistleblower, Spy: The Many Faces of Anonymous.* London: Verso.

Cooley, Mike. 1987. *Architect or Bee? The Human Price of Technology.* London: Hogarth Press.

Crabu, Stefano, Federica Giovanella, Leonardo Maccari, and Paolo Magaudda. 2015. "A Transdisciplinary Gaze on Wireless Community Networks." *Technoscienza* 6 (2): 113–134. http://www.tecnoscienza.net/index.php/tsj/article/view/238.

Croissant, Jennifer, and Laurel Smith-Doerr. 2008. "Organizational Contexts of Science: Boundaries and Relationships between University and Industry." In *The Handbook of Science and Technology Studies*, edited by Edward J. Hackett, Olga Amsterdamska, Michael Lynch, and Judy Wajcman, third edition, 691–718. Cambridge, MA: MIT Press.

D'Ignazio, Catherine, Alexis Hope, Alexandra Metral, David Raymond Willow Brugh, Becky Michelson, Tal Achituv, and Ethan Zuckerman. 2016. "Towards a Feminist Hackathon: The 'Make the Breast Pump Not Suck!' Hackathon." *Journal of Peer Production*, no. 6. http://peerproduction.net/issues/issue-8-feminism-and-unhacking-2/peer -reviewed-papers/towards-a-feminist-hackathon-the-make-the-breast-pump-not-suck/.

Dafermos, George. 2012. "Authority in Peer Production: The Emergence of Governance in the FreeBSD Project." *Journal of Peer Production*, no. 1. http://peerproduction .net/issues/issue-1/peer-reviewed-papers/authority-in-peer-production/.

Dafermos, George. 2012. *Governance Structures of Free/Open Source Software Development: Examining the Role of Modular Product Design as a Governance Mechanism in the FreeBSD Project*. NGInfra PhD Thesis Series on Infrastructures 51. Delft, The Netherlands: Next Generation Infrastructures Foundation.

Dagdelen, Demet. 2012. "Anonymous, WikiLeaks and Operation Payback: A Path to Political Action Through IRC and Twitter." Paper presented at the IPP2012: Big Data, Big Challenges?, Oxford Internet Institute, Oxford, UK. https://blogs.oii.ox.ac .uk/ipp-conference/sites/ipp/files/documents/Dagdelen2.pdf.

Dahlberg, Lincoln. 2015. "Which Social Media? A Call for Contextualisation." *Social Media + Society*. https://doi.org/10.1177/2056305115578142.

Dauvé, Gilles. 2011. "Communisation." Troploin. https://theanarchistlibrary.org/libr ary/gilles-dauve-communisation.

David, Paul A. 1985. "Clio and the Economics of QWERTY." *The American Economic Review* 75 (2): 332–337.

Davies, Sarah R. 2017. *Hackerspaces: Making the Maker Movement*. Cambridge: Polity.

De Filippi, Primavera, and Miguel Vieira. 2014. "The Commodification of Information Commons: The Case of Cloud Computing." *Columbia Science & Technology Law Review* 16 (1): 102–143.

Deleuze, Gilles, and Félix Guattari. 1987. *A Thousand Plateaus*. Minneapolis: University of Minnesota Press.

Delfanti, Alessandro. 2013. *Biohackers: The Politics of Open Science*. London: Pluto Press.

Delfanti, Alessandro. 2014. "Is Do-It-Yourself Biology Being Co-Opted by Institutions?" In *Meta Life: Biotechnologies, Synthetic Biology, A Life and the Arts*, edited by Annick Bureaud, Roger F. Malina, and Louise Whiteley. Leonardo Ebook Series. Cambridge, MA: MIT Press.

Delfanti, Alessandro. 2021. "Machinic Dispossession and Augmented Despotism: Digital Work in an Amazon Warehouse." *New Media & Society* 23 (1): 39–55. https://doi.org/10.1177/1461444819891613.

Delfanti, Alessandro, and Johan Söderberg. 2018. "Repurposing the Hacker: Three Cycles of Recuperation in the Evolution of Hacking and Capitalism." *Ephemera: Theory & Politics in Organization* 18 (3): 457–476. http://escholarship.org/uc/item/9c86493g.

Denker, Kai. 2014. "Heroes yet Criminals of the German Computer Revolution." In *Hacking Europe: From Computer Cultures to Demoscenes*, edited by Gerard Alberts and Ruth Oldenziel, first ed., 167–188. History of Computing. New York: Springer-Verlag.

de Zeeuw, Daniël. 2019. "Between Mass and Mask: The Profane Logic of Anonymous Imageboard Culture." PhD diss.: University of Amsterdam.

de Zeeuw, Daniël, Sal Hagen, Stijn Peeters, and Emilija Jokubauskaite. 2020. "Tracing Normiefication: A Cross-Platform Analysis of the QAnon Conspiracy Theory." *First Monday* 25 (11). https://doi.org/10.5210/fm.v25i11.10643.

Dickel, Sascha, and Jan-Felix Schrape. 2017. "The Logic of Digital Utopianism." *Nanoethics* 11: 47–58.

Dolata, Ulrich, and Jan-Felix Schrape. 2016. "Masses, Crowds, Communities, Movements: Collective Action in the Internet Age." *Social Movement Studies* 15 (1): 1–18.

Downing, John. 2001. *Radical Media: Rebellious Communication and Social Movements*. Thousand Oaks, CA: Sage.

Driscoll, Kevin. 2016. "Social Media's Dial-up Ancestor: The Bulletin Board System." *IEEE Spectrum*. http://spectrum.ieee.org/computing/networks/social-medias-dialup-ancestor-the-bulletin-board-system.

Dunbar-Hester, Christina. 2009. "Free the Spectrum! Activist Encounters with Old and New Media Technology." *New Media & Society* 11 (1–2): 221–240.

Dunbar-Hester, Christina. 2014. *Low Power to the People: Pirates, Protest, and Politics in FM Radio Activism*. Inside Technology. Cambridge, MA: MIT Press.

Dunbar-Hester, Christina. 2019. *Hacking Diversity: The Politics of Inclusion in Open Technology Cultures*. Princeton, NJ: Princeton University Press.

Dyer-Witheford, Nick. 2015. *Cyber-Marx: Cycles and Circuits of Struggle in High-Technology Capitalism*. Digital Barricades: Interventions in Digital Culture and Politics. London: Pluto Press.

Economist, The. 2011. "Print Me a Stradivarius." February 12, 2011.

Edgerton, David. 2008. *The Shock of the Old: Technology and Global History Since 1900*. London: Profile Books.

Ehn, Pelle. 1988. *Work-Oriented Design of Computer Artifacts*. First ed. Stockholm: Arbetslivcentrum. https://www.diva-portal.org/smash/get/diva2:580037/FULLTEXT02.

Ehn, Pelle, Elisabet M. Nilsson, and Richard Topgaard, eds. 2014. *Making Futures: Marginal Notes on Innovation, Design, and Democracy*. First ed. Cambridge, MA: MIT Press. https://mitpress.mit.edu/books/making-futures.

Ensmenger, Nathan, and William Aspray. 2002. "Software as Labour Process." In *History of Computing: Software Issues*, edited by Ulf Hashagen, Reinhard Keil-Slawik, and Arthur L. Norberg, 139–165. Berlin: Springer-Verlag. https://doi.org/10.1007/978-3-662-04954-9.

Enzensberger, Hans Magnus. 1970. "Constituents of a Theory of the Media." *New Left Review* 64 (November–December).

Escamot Espiral. 2008. "Escamot Espiral." Website (archived). http://web.archive.org/web/20120621100513/http://kernelpanic.hacklabs.org/ee/index2.html.

European Commission Directorate-General for the Information Society and Media. 2009. *14th Progress Report on the Single European Electronic Communications Market: Final Version*. Brussels: European Commission.

"A Fablab Burned Down in France by Anarchists." 2017. Editorial, *Makery*, November 11, 2017. https://www.makery.info/en/2017/11/28/apres-lincendie-de-la-casemate-la-communaute-des-fablabs-reagit/.

Fichtner, Mirko "macro," ed. 2015. *C-Booc: 20 Years of c-Base*. First ed. Berlin: c-base. https://www.kickstarter.com/projects/macrone/c-booc-20-years-of-c-base-in-a-book/posts/1320184.

Fish, Adam, and Luca Follis. 2019. *Hacker States*. Cambridge, MA: MIT Press.

Flichy, Patrice. 2007. *The Internet Imaginaire*. Cambridge, MA: MIT Press.

Florida, Richard. 2005. *Cities and the Creative Class*. London: Routledge.

Flowers, Stephen. 2008. "Harnessing the Hackers: The Emergence and Exploitation of Outlaw Innovation." *Research Policy* 37 (2): 177–193. https://doi.org/10.1016/j.respol.2007.10.006.

Foster, Ellen K., and Yana Boeva. 2018. "Making Sense of Place: Research Reflections from Two Multi-Sited Ethnographies." *Making Futures* 5.

Frechette, Ian, and Helen Rose. 2007. "Early IRC History." Web page. http://www.efnet.org/?module=docs&doc=22.

Friedman, Andrew. 1977. *Industry and Labour: Class Struggle at Work and Monopoly Capitalism*. London: Macmillan.

Friends of the Classless Society. 2016. "On Communisation and Its Theorists." Endnotes. https://endnotes.org.uk/posts/friends-of-the-classless-society-on-communisation-and-its-theorists.

Fuchs, Christian. 2014. *Digital Labour and Karl Marx*. London: Routledge.

Fuchs, Christian. 2015a. "Dallas Smythe Today: The Audience Commodity, the Digital Labour Debate, Marxist Political Economy and Critical Theory. Prolegomena to a Digital Labour Theory of Value." In *Marx and the Political Economy of the Media*, edited by Christian Fuchs and Vincent Mosco, Vol. 79: 522–599. Studies in Critical Social Sciences. Leiden: Brill. https://doi.org/10.1163/9789004291416_019.

Fuchs, Christian. 2015b. "The Digital Labour Theory of Value and Karl Marx in the Age of Facebook, YouTube, Twitter and Weibo." In *Reconsidering Value and Labour in the Digital Age*, edited by Eran Fisher and Christian Fuchs, 26–41. London: Palgrave-Macmillan. https://doi.org/10.1057/978113747857.

Gallegos, Jenna E., Christopher Boyer, Eleanor Pauwels, Warren E. Kaplan, and Jean Peccoud. 2018. "The Open Insulin Project: A Case Study for 'Biohacked' Medicines." *Cell Press Reviews* 36 (12): 1211–1218. https://doi.org/10.1016/j.tibtech.2018.07.009.

Gershenfeld, Neil A. 2005. *Fab: The Coming Revolution on Your Desktop—from Personal Computers to Personal Fabrication*. New York: Basic Books.

Gill, Rosalind, and Andy Pratt. 2008. "In the Social Factory? Immaterial Labour, Pre-cariousness and Cultural Work." *Theory, Culture & Society* 25 (7–8): 1–30. https://doi.org/10.1177/0263276408097794.

Gillespie, Tarleton. 2006. "Engineering a Principle: 'End-to-End' in the Design of the Internet." *Social Studies of Science* 36 (3): 427–457. https://doi.org/10.1177/03063217 06056047.

Graham, Paul. 1993. *On Lisp: Advanced Techniques for Common Lisp*. First ed. Upper Saddle River, NJ: Prentice Hall.

Graham, Paul. 2004a. *Hackers & Painters: Big Ideas from the Computer Age*. Sebastopol, CA: O'Reilly Media.

Graham, Paul. 2004b. "The Word 'Hacker.'" Blog entry. http://www.paulgraham.com /gba.html.

Graham, Paul. 2005. "Summer Founders Program [Announcement]." Blog post (archived), March, 2005. http://web.archive.org/web/20150205091134/http://paulgraham.com /summerfounder.html.

Graham, Paul. 2012. "How Y Combinator Started." Y Combinator (archived), March 15, 2012. http://web.archive.org/web/20180311213728/http://old.ycombinator.com /start.html.

Graham, Stephen, and Simon Marvin. 2001. *Splintering Urbanism: Networked Infra-structures, Technological Mobilities and the Urban Condition*. London: Routledge.

Gramsci, Antonio. 1971. *Selections from the Prison Notebooks*. New York: International Publishers.

Greater London Enterprise Board. 1984. *Technology Networks: Science and Technology Serving London's Needs*. London: Greater London Enterprise Board.

"Grenoble Technopole Apaisée?" 2016. Article on Nantes Indymedia. https://nantes .indymedia.org/articles/39247.

Grier, David Alan, and Mary Campbell. 2000. "A Social History of Bitnet and Listserv 1985–1991." *IEEE Annals of the History of Computing* 22 (2): 32–41. https://ieeexplore .ieee.org/iel5/85/18190/00841135.pdf.

Hackerspaces Wiki contributors. 2020a. "Hackerspaces.org." Landing page. https:// hackerspaces.org/.

Hackerspaces Wiki contributors. 2020b. "List of Currently Active Hackerspaces." Wiki page. https://wiki.hackerspaces.org/List_of_hackerspaces.

Hampton, Keith N., and Neeti Gupta. 2008. "Community and Social Interaction in the Wireless City: Wi-Fi Use in Public and Semi-Public Spaces." *New Media & Society* 10 (6): 831–850. https://doi.org/10.1177/1461444808096247.

Hanlon, Gerard. 2016. *The Dark Side of Management: A Secret History of Management Knowledge*. London: Routledge.

Harvey, David. 2005. *A Brief History of Neoliberalism*. Oxford: Oxford University Press.

Hayek, Friedrich A. 1948. "The Use of Knowledge in Society." In *Individualism and Economic Order*. First ed., 77–91. Chicago: University of Chicago Press.

Helmond, Anne. 2015. "The Platformization of the Web: Making Web Data Platform Ready." *Social Media + Society* 1 (2).

Helmond, Anne, David B. Nieborg, and Fernando N. van der Vlist. 2019. "Facebook's Evolution: Development of a Platform-as-Infrastructure." *Internet Histories* 3 (2): 123–146. https://doi.org/10.1080/24701475.2019.1593667.

Hess, David J. 2005. "Technology- and Product-Oriented Movements: Approximating Social Movement Studies and STS." *Science, Technology and Human Values* 30 (4): 515–535. https://doi.org/10.1177/0162243905276499.

Himanen, Pekka. 2001. *The Hacker Ethic*. New York: Random House.

Hodgson, Gary. 2013. "Interview with Nophead." *RepRap Magazine*, no. 2: 75–78. https://xyzdims.com/wp-content/uploads/2018/07/reprap-magazine-21.pdf.

Hogan, Bernie, and Anabel Quan-Haase. 2010. "Persistence and Change in Social Media." *Bulletin of Science, Technology & Society* 4: 309–315.

Horst, Maja, and Alan Irwin. 2010. "Nations at Ease with Radical Knowledge: On Consensus, Consensusing and False Consensusness." *Social Studies of Science*. 40 (1): 105–126.

Hurst, Nathan. 2014. "TechShop's Not-so-Secret Ingredient: To Achieve its Prodigious Goals, the Marquee Makerspace is Dialing in a Precise Method to Make Each New Shop Polished and Profitable." *Make: Magazine* (archived) 40: 54–58. http://web .archive.org/web/20161221170306/https://p25ext.lanl.gov/people/hubert/outreach /make/M40_spaces.pdf.

Ikemoto, Lisa C. 2017. "DIY Bio: Hacking Life in Biotech's Backyard." *U.C.D. Law Review* 51: 539–568. https://lawreview.law.ucdavis.edu/issues/51/2/Symposium/51-2_Ikemoto .pdf.

Irani, Lilly. 2013. "The Cultural Work of Microwork." *New Media & Society* 17 (5): 720–739. https://doi.org/10.1177/1461444813511926.

Irani, Lilly. 2015a. "Difference and Dependence Among Digital Workers: The Case of Amazon Mechanical Turk." *The South Atlantic Quarterly* 114 (1): 225–234. https://doi.org/10.1215/00382876-2831665.

Irani, Lilly. 2015b. "Hackathons and the Politics of Entrepreneurial Time." *Science, Technology and Human Values* 40 (5): 799–824. https://doi.org/10.1177/0162243915578486.

Jenkins, Henry, Sam Ford, and Joshua Green. 2013. "What Constitutes Meaningful Participation?" In *Spreadable Media: Creating Value and Meaning in a Networked Culture*, 153–194. New York: NYU Press.

Jessop, Bob. 1995. "The Regulation Approach, Governance and post-Fordism: Alternative Perspectives on Economic and Political Change?" *Economy and Society* 24 (3): 307–333. https://doi.org/10.1080/03085149500000013.

Jordan, Tim. 2016. "A Genealogy of Hacking." *Convergence: The International Journal of Research into New Media Technologies* 23 (5): 528–544. https://doi.org/10.1177/1354856516640710.

Juris, Jeffrey S. 2005. "The New Digital Media and Activist Networking Within Anti-Corporate Globalization Movements." *The ANNALS of the American Academy of Political and Social Science* 597 (1): 189–208. https://doi.org/10.1177/000271620 4270338.

Keane, Michael, and Elaine Jing Zhao. 2012. "Renegades on the Frontier of Innovation: The Shanzhai Grassroots Communities of Shenzhen, China." *Eurasian Geography and Economics* 53 (2): 216–230. https://doi.org/10.2747/1539-7216.53.2.216.

Kelty, Christopher. 2005. "Geeks, Social Imaginaries, and Recursive Publics." *Cultural Anthropology* 20 (2): 185–214.

Kelty, Christopher. 2008. *Two Bits: The Cultural Significance of Free Software*. Durham, NC: Duke University Press. http://twobits.net/.

Kelty, Christopher. 2019. "Participation, Administered." In *The Participant: A Century of Participation in Four Stories*, 136–182. Chicago: University of Chicago Press.

Kirkpatrick, Graeme. 2018. *Critical Technology: A Social Theory of Personal Computing*. Reissued. London: Routledge.

Kleif, Tine, and Wendy Faulkner. 2003. "'I'm No Athlete [but] I Can Make This Thing Dance!': Men's Pleasures in Technology." *Science, Technology, & Human Values* 28 (2): 296–325. https://doi.org/10.1177/0162243902250908.

Kohtala, Cindy, and Camille Bosqué. 2014. "The Story of MIT-Fablab Norway: A Narrative on Infrastructuring Peer Production." *Journal of Peer Production*, no. 5.

Kostakis, Vasilis, Vasilis Niaros, and Christos Giotitsas. 2015. "Production and Governance in Hackerspaces: A Manifestation of Commons-Based Peer Production in the Physical Realm?" *International Journal of Cultural Studies* 18 (2): 555–573. https://doi.org/10.1177/136877913519310.

Kovel, Joel. 2008. "Dialectic as Praxis." In *Dialectics for the Twenty-First Century*, edited by Bertell Ollman and Tony Smith, 235–242. New York: Palgrave Macmillan.

Kraft, Phillip, and Jørgen P. Bansler. 1994. "The Collective Resource Approach: The Scandinavian Experience." *Scandinavian Journal of Information Systems* 6 (1): 71–84.

Kubitschko, Sebastian. 2015. "The Role of Hackers in Countering Surveillance and Promoting Democracy." *Media and Communication* 3 (2): 77–87. https://doi.org/10.17645/mac.v3i2.281.

Kuhn, Thomas S. 1962. *The Structure of Scientific Revolutions*. Chicago: University of Chicago Press.

Kurlander, David, Tim Skelly, and David Salesin. 1996. "Comic Chat." In *SIGGRAPH '96: Proceedings of the 23rd Annual Conference on Computer Graphics and Interactive Techniques*, edited by John Fujii, 225–236. New York: ACM. https://doi.org/10.1145/237170.237260.

Lakhani, Karim R., and Robert Wolf. 2005. "Why Hackers Do What They Do: Understanding Motivation and Effort in Free/Open Source Software Projects." In *Perspectives on Free and Open Source Software*, edited by Joe Feller, Brian Fitzgerald, Scott Hissam, and Karim R. Lakhani. Cambridge, MA: MIT Press.

Landers, Chris. 2008. "Serious Business: Anonymous Takes on Scientology (and Doesn't Afraid of Anything)." News article in *Baltimore City Paper*. http://www.chrislanders.net/serious-business/.

Lapsley, Phil, and Steve Wozniak. 2013. *Exploding the Phone: The Untold Story of the Teenagers and Outlaws Who Hacked Ma Bell*. New York: Grove Press. http://explodingthephone.com/.

Latzko-Toth, Guillaume. 2010. "Metaphors of Synchrony: Emergence Differentiation of Online Chat Devices." *Bulletin of Science, Technology & Society* 30 (5): 362–374. https://doi.org/10.1177/0270467610380005.

Latzko-Toth, Guillaume. 2013. "A Tale of Two IRC Networks." In *Qualitative Research: The Essential Guide to Theory and Practice*, edited by Maggi Savin-Baden and Claire H. Major, 209–210. London: Routledge.

Latzko-Toth, Guillaume, and Maxigas. 2019. "An Obscure Object of Communicational Desire: The Untold Story of Online Chat." In *Second International Handbook of Internet Research*, edited by Jeremy Hunsinger, Matthew M. Allen, and Lisbeth Klastrup, 381–394. Dordrecht, The Netherlands: Springer. https://doi.org/10.1007/978-94-024-1555-1_8.

Layton, Edwin. 1986. *The Revolt of the Engineers: Social Responsibility and the American Engineering Profession*. Baltimore: Johns Hopkins University Press.

Lee, Matt. 2008. "El Software Libre En La Etica y La Practica [Free Software in Ethics and Practice]." Event announcement on the website of the Free Software Foundation. https://www.fsf.org/events/20080409malaga.

Lee, Yu-Hao, and Holin Lin. 2011. "'Gaming Is My Work': Identity Work in Internet-Hobbyist Game Workers." *Work, Employment & Society* 25 (3): 451–467. https://doi.org/10.1177/0950017011407975.

Leistert, Oliver. 2016. "Social Bots as Algorithmic Pirates and Messengers of Techno-Environmental Agency." In *Algorithmic Cultures: Essays on Meaning, Performance and New Technologies*, edited by Jonathan Roberge and Robert Seyfert, first ed., 158–172. Routledge Advances in Sociology 189. London: Routledge.

Levy, Steven. 1984. *Hackers: Heroes of the Computer Revolution*. New York: Anchor Press, Doubleday.

Leydesdorff, Loet, and Peter Van Den Besselaar. 1987. "What We Have Learned from the Amsterdam Science Shop." In *The Social Direction of the Public Sciences*, edited by Stuart Blume, Joske Bunders, Loet Leydesdorff, and Richard Whitley, 11: 135–162. Sociology of the Sciences Yearbook. Dordrecht, The Netherlands: D. Reidel Publishing Company. https://doi.org/10.1007/978-94-009-3755-0.

Lindtner, Silvia. 2015. "Hacking with Chinese Characteristics: The Promises of the Maker Movement against China's Manufacturing Culture." *Science, Technology & Human Values* 40 (5): 854–879.

Lindtner, Silvia, and David Li. 2012. "Created in China: The Makings of China's Hackerspace Community." *Interactions* 19 (6): 18–22. https://doi.org/10.1145/2377783.2377789.

Lindtner, Silvia, Garnet Hertz, and Paul Dourish. 2014. "Emerging Sites of HCI Innovation: Hackerspaces, Hardware Startups & Incubators." *ACM Transactions on Computer-Human Interaction*. http://www.wiwi.uni-siegen.de/inno/pdf/lindtner_maker chinainnovation_chi14.pdf.

Liu, Alan. 2004. *The Laws of Cool*. Chicago: University of Chicago Press.

López, Miguel A. Martínez. 2013. "The Squatters' Movement in Europe: A Durable Struggle for Social Autonomy in Urban Politics." *Antipode* 45 (4): 866–887. https://doi.org/10.1111/j.1467-8330.2012.01060.x.

Luque-Ayala, Andrés, and Simon Marvin. 2020. *Urban Operating Systems: Producing the Computational City*. Infrastructure Series. Cambridge, MA: MIT Press.

Marx, Karl. 1994. "Results of the Direct Production Process." In *Marx: 1861–1864*. Marx & Engels Collected Works, Vol. 34. London: Lawrence & Wishart.

Mauss, Marcel. 2016. *The Gift*. Translated by Jane I. Guyer. Expanded edition. Chicago: HAU.

Mauvaise Troupe. 2014. "Hacker Vaillants." In *Constellations: Trajectories Révolutionnaires Du Jenue 21è Siècle*. https://mauvaisetroupe.org/spip.php?rubrique1.

Maxigas. 2012. "Hacklabs and Hackerspaces: Tracing Two Genealogies." *Journal of Peer Production* 2. http://peerproduction.net/issues/issue-2/peer-reviewed-papers/hacklabs-and-hackerspaces/.

Maxigas. 2013. "Only a Committee Can Save Us." Presentation at esCTS (Annual Conference of the Spanish Network of STS Researchers). June 19, 2013, Barcelona. http://redescts.files.wordpress.com/2013/05/programa.pdf.

Maxigas. 2014. "Cultural Stratigraphy: A Historical Rift in the Hacker Scene Between Hacklabs and Hackerspaces." *Journal of Peer Production*, no. 5. http://peerproduction.net/issues/issue-5-shared-machine-shops/editorial-section/cultural-stratigraphy-a-rift-between-shared-machine-shops/.

Maxigas. 2015. "Hacklabs and Squats: Engineering Counter-Cultures in Autonomous Spaces." In *Making Room: Cultural Production in Occupied Spaces*, edited by Alan Moore and Alan Smart, 328–341. Los Angeles: Other Forms & the Journal of Aesthetics and Protest. http://joaap.org/press/makingroom.htm.

Maxigas. 2015. "Peer Production of Open Hardware: Unfinished Artefacts and Architectures in the Hackerspaces." PhD diss., Universitat Oberta de Catalunya, Internet Interdisciplinary Institute.

Maxigas. 2017. "Keeping Technological Sovereignty: The Case of Internet Relay Chat." In *Technological Sovereignty 2*, edited by Alex Haché. Vol. 2. Dossier Ritimo. Paris: Ritimo.

Maxigas, and Guillaume Latzko-Toth. 2020. "Trusted Commons: Why 'Old' Social Media Matter." *Internet Policy Review* 9 (4). https://policyreview.info/articles/analysis/trusted-commons-why-old-social-media-matter.

Mirowski, Philip. 2011. "Pharma's Market: New Horizons in Outsourcing in the Modern Globalized Regime." In *Science-Mart: Privatizing American Science*, 194–258. Cambridge, MA: Harvard University Press.

Moilanen, Jarkko, aka kyb3R. 2010. "Hackerspaces, Members and Involvement (Survey Study)." Blog entry. http://extreme.ajatukseni.net/2010/07/19/hackerspaces-members-and-involvement-survey-study/.

Montgomery, David. 1976. "Workers' Control of Machine Production in the Nineteenth Century." *Labor History* 17 (4): 485–509.

Montgomery, David. 1987. *The Fall of the House of Labor: The Workplace, the State, and American Labor Activism, 1865–1925*. Cambridge: Cambridge University Press.

Moody, Glyn. 2001. *Rebel Code: Linux and the Open Source Revolution*. New York: Perseus Publishing.

Moorhouse, H. F. 1978. "The Marxist Theory of the Labour Aristocracy." *Social History* 3 (1): 61–82. https://doi.org/10.1080/03071027808567419.

Morris, William. 1988. "Chapter 4: The Voice of John Ball." In *A Dream of John Ball*. London: Reeves and Turner.

Mota, Sofia Catarina Mósca Ferreira. 2014. "Bits, Atoms, and Information Sharing: New Opportunities for Participation." PhD diss., Universidade Nova de Lisboa. http://hdl.handle.net/10362/14505.

Mulinari, Shai, Andreas Vilhelmsson, Rickard Emily, and Piotr Ozieranski. 2020. "Five Years of Pharmaceutical Industry Funding of Patient Organisations in Sweden: Cross-Sectional Study of Companies, Patient Organisations and Drugs." *PLOS ONE*. https://doi.org/10.1371/journal.pone.0235021.

Murillo, Luis Felipe R. 2019. "Hackerspaces Network: Prefiguring Technopolitical Futures?" *American Anthropologist* 122 (2): 207–221. https://doi.org/10.1111/aman.13318.

Mutton, Paul. 2004. *IRC Hacks: 100 Industrial-Strength Tips & Tools*. Cambridge, MA: O'Reilly Media.

Nagle, Angela. 2017. *Kill All Normies: Online Culture Wars from 4Chan and Tumblr to Trump and the Alt-Right*. Winchester, UK: Zero Books.

Negri, Antonio. 1989. *The Politics of Subversion: A Manifesto for the Twenty-First Century*. Cambridge: Polity Press.

Noble, David. 1977. *America by Design: Science, Technology, and the Rise of Corporate Capitalism*. First ed. New York: Knopf.

Ohlig, Jens, and Lars Weiler. 2007. "Building a Hackerspace." Talk at the Twenty-fourth Chaos Communication Congress (24C3), December 2007, Berlin.

Oliver, Julian, Gordan Savičić, and Danja Vasiliev. 2011. "The Critical Engineering Manifesto." Web page. https://criticalengineering.org/.

Ostrom, Elinor, Joanna Burger, Christopher B. Field, Richard B. Norgaard, and David Policansky. 1999. "Revisiting the Commons: Local Lessons, Global Challenges." *Science* 284 (278): 278–282. https://doi.org/10.1126/science.284.5412.278.

Pansa, Giampaolo. 2007. "Fiat Has Branded Me." In *Autonomia: Post-Political Politics*, edited by Sylvére Lotringer and Christian Marazzi, 24–27. Los Angeles: Semiotext(e).

Picon, Antoine. 2009. "The Engineer as Judge: Engineering Analysis and Political Economy in Eighteenth Century France." *Engineering Studies* 1 (1): 19–34. https://doi.org/10.1080/19378620902725174.

Plantin, Jean-Chrisophe, Carl Lagoze, Paul N. Edwards, and Christian Sandvig. 2018. "Infrastructure Studies Meet Platform Studies in the Age of Google and Facebook." *New Media & Society* 20 (1): 293–310. https://doi.org/10.1177/1461444816661553.

Postigo, Hector. 2004. "Emerging Sources of Labor on the Internet: The Case of America Online Volunteers." In *Uncovering Labour in Information Revolutions, 1750–2000*, edited by Aad Blok and Greg Downey, 205–223. London: Cambridge University Press.

Powell, Alison. 2012. "Democratizing Production through Open Source Knowledge: From Open Software to Open Hardware." *Media, Culture & Society* 34 (6): 691–708. https://doi.org/10.1177/0163443712449497.

Pynchon, Thomas. 2013. *Bleeding Edge*. London: Penguin.

Rajan, Kaushik. 2017. *Pharmocracy: Value, Politics, and Knowledge in Global Biomedicine*. London: Duke University Press.

Rapatzikou, Tatiani G. 2004. *Gothic Motifs in the Fiction of William Gibson*. Postmodern Studies 23. New York: Rodopi.

Raymond, Eric S. 1999. *The Cathedral and the Bazaar: Musings on Linux and Open Source by an Accidental Revolutionary*. Sebastopol, CA: O'Reilly. http://www.catb.org/textasciitildeesr/writings/cathedral-bazaar/cathedral-bazaar/.

Remneland-Wikhamn, Björn, Jan Ljungberg, Magnus Bergquist, and Jonas Kuschel. 2011. "Open Innovation, Generativity and the Supplier as Peer: The Case of iPhone and Android." *International Journal of Innovation Management* 15 (1). https://doi.org/10.1142/S1363919611003131.

Renee, Marlin-Bennet. 2017. "Science in Whose Interest? States, Firms, the Public, and Scientific Knowledge." In *Who Owns Knowledge? Knowledge and the Law*, edited by Bernd Weiler, 125–152. New York; London: Routledge.

Rigi, Jakob. 2012. "Peer to Peer Production and Advanced Communism: The Alternative to Capitalism." *Journal of Peer Production* 1. http://peerproduction.net/issues/issue-1/invited-comments/a-new-communist-horizon/.

Rigi, Jakob. 2013. "Peer Production and Marxian Communism: Contours of a New Emerging Mode of Production." *Capital & Class* 37 (3): 397–416. https://doi.org/10.1177/0309816813503979.

"'River Rat' Says Truckers Forgotten." 1975. Cape Girardeau Southeast Missourian, Sunday, February 16, 1975. https://news.google.com/newspapers?id=qm0fAAAAIBAJ&sjid=ZtUEAAAAIBAJ&pg=841%2C5104091.

Rogers, Richard. 2013. *Digital Methods*. Cambridge, MA: MIT Press.

Rogers, Richard. 2020. "Deplatforming: Following Extreme Internet Celebrities to Telegram and Alternative Social Media." *European Journal of Communication* 35 (3): 213–229. https://doi.org/10.1177/0267323120922066.

Rosich, Gerard. 2019. *The Contested History of Autonomy*. London: Bloomsbury.

Rowland, Wade. 2005. "Recognizing the Role of the Modern Business Corporation in the 'Social Construction' of Technology." *Social Epistemology* 19 (2–3): 287–313. https://doi.org/10.1080/02691720500145522.

Russell, Andrew L. 2014. *Open Standards and the Digital Age: History, Ideology, and Networks*. First ed. Cambridge Studies in the Emergence of Global Enterprise. Cambridge: Cambridge University Press.

Sadofsky, Jason Scott. 2005. "BBS: The Documentary." Documentary.

Sandvig, Christian. 2004. "An Initial Assessment of Cooperative Action in Wi-Fi Networking." *Telecommunications Policy* 28 (7–8): 579–602. https://doi.org/10.1016 /j.telpol.2004.05.006.

Sauter, Molly. 2014. *The Coming Swarm: DDoS, Hacktivism, and Civil Disobedience.* New York: Bloomsbury.

Schneider, David. 1998. "The Innovators Club: Interest in TechShop's Neighborhood Workshops Is Growing." *IEEE Spectrum* 45 (10): 22–23. https://doi.org/10.1109 /MSPEC.2008.4635045.

Scholz, Trebor. 2016. *Uberworked and Underpaid: How Workers Are Disrupting the Digital Economy.* Cambridge: Polity Press.

Senker, Peter, and Mark Beesley. 1986. "The Need for Skills in the Factory of the Future." *New Technology, Work and Employment* 1 (1): 9–17. https://doi.org/10.1111/j .1468-005X.1986.tb00076.x.

Seravalli, Anna. 2012. "Infrastructuring for Opening Production, from Participatory Design to Participatory Making?" In *Proceedings of the Twelfth Participatory Design Conference: Exploratory Papers, Workshop Descriptions, Industry Cases—Volume 2,* 53–56. PDC 2012. New York: ACM. https://doi.org/10.1145/2348144.2348161.

Sharlin, Harold Issadore. 1976. "Herbert Spencer and Scientism." *Annals of Science* 33 (5): 457–480. https://doi.org/10.1080/00033797600200641.

Shirky, Clay. 2008. *Here Comes Everybody: The Power of Organizing without Organizations.* New York: Penguin.

Smith, Adrian. 2014. "Technology Networks for Socially Useful Production." *Journal of Peer Production,* no. 5. https://core.ac.uk/download/pdf/30611280.pdf.

Smith, Adrian, Mariano Fressoli, Dinesh Abrol, Elisa Arond, and Adrian Ely. 2017. *Grassroots Innovation Movements.* First ed. Pathways to Sustainability. New York: Routledge.

Smith, Ernie. 2021. "The Default Router: How Linksys' most Famous Router, the Wrt54g, Tripped into Legendary Status because of an Undocumented Feature that Slipped through during a Merger." Tedium. https://tedium.co/2021/01/13/linksys -wrt54g-router-history/.

Smythe, Dallas W. 1977. "Communications: Blindspot of Western Marxism." *Canadian Journal of Political and Society Theory* 1 (3): 1–28.

Söderberg, Johan. 2010. "Reconstructivism versus Critical Theory of Technology: Alternative Perspectives on Activism and Institutional Entrepreneurship in the Czech Wireless Community." *Social Epistemology: A Journal of Knowledge, Culture and Policy* 24 (4): 239–262. https://doi.org/10.1080/02691728.2010.506962.

Söderberg, Johan. 2011. "Free Software to Open Hardware: Critical Theory on the Frontiers of Hacking." PhD diss., University of Gothenburg. https://gupea.ub.gu.se /bitstream/2077/24450/4/gupea_2077_24450_4.pdf.

Söderberg, Johan. 2013. "Automating Amateurs in the 3D Printing Community: Connecting the Dots Between 'Deskilling' and 'User-Friendliness.'" *Work Organisation, Labour & Globalisation* 7 (1): 124–139.

Söderberg, Johan. 2013. "Determining Social Change: The Role of Technological Determinism in the Collective Action Framing of Hackers." *New Media & Society* 15 (8): 1277–1293. https://doi.org/10.1177/1461444812470093.

Söderberg, Johan. 2014. "Reproducing Wealth without Money, One 3D Printer at a Time: The Cunning of Instrumental Reason." *Journal of Peer Production,* no. 4. http://peerproduction.net/issues/issue-4-value-and-currency/peer-reviewed-articles /reproducing-wealth-without-money/.

Söderberg, Johan. 2019. "The Cloud Factory: Making Things and Making a Living with Desktop 3D Printing." *Culture and Organization* 25 (1): 65–81. https://doi.org/10 .1080/14759551.2016.1203313.

Söderberg, Johan, and Adel Daoud. 2012. "Atoms Want to Be Free Too! Expanding the Critique of Intellectual Property to Physical Goods." *TripleC: Cognition, Communication, Co-Operation* 10 (1). http://www.triple-c.at/index.php/tripleC/article /view/288.

Soufron, Jean-Baptiste. 2017. "In France, Cyber Criticism Turns Violent as 'Activists' Burn a Fablab to Protest the Diffusion Of . . ." *Hackernoon.*https://hackernoon .com/in-france-cyber-criticism-turns-violent-as-activists-burn-a-fablab-to-protest-the -diffusion-of-4ad378251c5b.

Srnicek, Nick. 2016. *Platform Capitalism.* London: Polity Press.

Stallman, Richard M. 1993. "The GNU Manifesto." Web page. https://www.gnu.org /gnu/manifesto.en.html.

Stallman, Richard M. 2002. *Free Software, Free Society: Selected Essays of Richard M. Stallman.* Edited by Joshua Gay. Boston: GNU Press.

Stallman, Richard M. 2015. "Bill Gates and Other Communists." Article on the Free Software Foundation website, originally published on CNET in 2005. https://www .gnu.org/philosophy/bill-gates-and-other-communists.en.html.

Steele, Guy L., and Eric S. Raymond, eds. 1996. *The New Hacker's Dictionary.* Third ed. Cambridge, MA: MIT Press.

Stenberg, Daniel. 2011. "History of IRC (Internet Relay Chat)." Web page. http:// daniel.haxx.se/irchistory.html.

"storia." 2010. HackIt 0x0C website (archived). http://web.archive.org/web/20100613 015928/http://hackmeeting.org/hackit09/index.php?page=storia&lang=en.

Tapscott, Don, and Anthony D. Williams. 2006. *Wikinomics: How Mass Collaboration Changes Everything.* New York: Penguin.

Terranova, Tiziana. 2000. "Free Labor: Producing Culture for the Digital Economy." *Social Text* 18 (2): 33–58.

Thomas, Douglas. 2002. *Hacker Culture*. Minneapolis: University of Minnesota Press.

Thomas, Greg. 2014. "How a German Soda Became Hackers' Fuel of Choice." Article in *Vice Magazine*. http://www-refresh.vice-motherboard-test.appspot.com/blog/how -a-german-soda-became-hackers-fuel-of-choice.

Thompson, E. P. 1963. *The Making of the English Working Class*. London: Victor Gollancz. "The Thoughtless Information Technologist." 1966. *Datamation* 12 (8).

Tocchetti, Sara, and Sara Angeli Aguiton. 2015. "Is an FBI Agent a DIY Biologist Like Any Other? A Cultural Analysis of a Biosecurity Risk." *Science, Technology, & Human Values* 40 (5): 825–853. https://doi.org/10.1177/0162243915589634.

Toupin, Sophie. 2013. "Feminist Hackerspaces as Safer Spaces?" *dpi: Feminist Journal of Art and Digital Culture*, no. 27. http://dpi.studioxx.org/en/feminist-hackerspaces-safer -spaces.

Toupin, Sophie. 2014. "Feminist, Queer and Trans Hackerspaces: The Crystallization of an Alternate Hacker Culture?" *Journal of Peer Production*, no. 5. http://peerproduction .net/issues/issue-5-shared-machine-shops/peer-reviewed-articles/feminist-hackerspaces -the-synthesis-of-feminist-and-hacker-cultures/.

Tronti, Mario. 1979. "Lenin in England." In *Working Class Autonomy and the Crisis: Italian Marxist Texts of the Theory and Practice of a Class Movement, 1964–79*, edited by Red Notes Collective, 1–6. London: Red Notes/CSE Books.

Troxler, Peter. 2015. "Fab Labs Forked: A Grassroots Insurgency Inside the Next Industrial Revolution." *Journal of Peer Production*, no. 5. http://peerproduction.net /issues/issue-5-shared-machine-shops/editorial-section/fab-labs-forked-a-grassroots -insurgency-inside-the-next-industrial-revolution/.

Troxler, Peter, and Maxigas. 2014. "We Now Have the Means of Production, but Where Is My Revolution?" *Journal of Peer Production*, no. 5. http://peerproduction .net/issues/issue-5-shared-machine-shops/editorial-section/editorial-note-we-now -have-the-means-of-production-but-where-is-my-revolution/.

Turner, Fred. 2006. *From Counterculture to Cyberculture: Stewart Brand, the Whole Earth Network, and the Rise of Digital Utopianism*. First ed. Chicago: University of Chicago Press.

Túry, György. 2015. "Leftist vs. (Neo)liberal Script for the (Media) Future: Enzens-berger's 'Constitutents of a Theory of the Media.'" *International Journal of Cultural Studies* 18 (6): 613–627. https://doi.org/10.1177/136787791454473.

van der Nagel, Emily. 2017. "From Usernames to Profiles: The Development of Pseudonymity in Internet Communication." *Internet Histories* 1 (4): 312–331. https:// doi.org/10.1080/247014575.2017.1389548.

van Doorn, Niels. 2011. "Digital Spaces, Material Traces: How Matter Comes to Matter in Online Performances of Gender, Sexuality and Embodiment." *Media, Culture & Society* 33 (4): 531–547. https://doi.org/10.1177/0163443711398692.

van Oost, Ellen, Stefan Verhaegh, and Nelly Oudshoorn. 2009. "From Innovation Community to Community Innovation: User-Initiated Innovation in Wireless Leiden." *Science, Technology, & Human Values* 34 (2): 182–205. https://doi.org/10.1177 /0162243907311556.

Veblen, Thorstein. 2001. *The Engineers and the Price System.* Kitchener, Canada: Batoche.

von Hippel, Eric. 2005. *Democratizing Innovation.* Cambridge, MA: MIT Press. http:// web.mit.edu/people/evhippel/democ1.htm.

von Hippel, Eric. 2016. *Free Innovation.* Cambridge, MA: MIT Press.

Wark, McKenzie. 2004. *A Hacker Manifesto.* First ed. Cambridge, MA: Harvard University Press.

Weber, Steven. 2004. *The Success of Open Source.* Cambridge, MA: Harvard University Press.

Wenten, Klara-Aylin. 2019. "Controlling Labor in Makeathons: On the Recuperation of Emancipation in Industrial Labor Processes." In *Digitalization in Industry: Between Domination and Emancipation*, edited by Uli Meyer, Simon Schaupp, and David Seibt, 153–178. Cham, Switzerland: Palgrave-Macmillan. https://doi.org/10.1007/978-3 -030-28258-5.

Whatley, John. 2013. "Gothic Self-Fashioning in Gibson's Novels: Nature, Culture, Identity, Improvisation, and Cyberspace." In *A Companion to American Gothic*, edited by Charles L. Crow, first ed., 418–433. Blackwell Companions to Literature and Culture. Chichester, UK: Wiley Blackwell.

Williams, Rosalind. 2003. *Retooling: A Historian Confronts Technological Change.* Cambridge, MA: MIT Press.

Wilson, Nathan J., and Reinie Cordier. 2013. "A Narrative Review of Men's Sheds Literature: Reducing Social Isolation and Promoting Men's Health and Well-Being." *Health and Social Care in the Community* 21 (5): 451–463. https://doi.org/10.1111/hsc.12019.

Wood, David Murakami, and Torin Mohanan. 2019. "Platform Surveillance." *Surveillance & Society* 17 (1/2): 1–6.

Wood, Steve. 1987. "The Deskilling Debate, New Technology, and Work Organization." *Acta Sociologica* 30 (1): 3–24. https://doi.org/10.1177/000169938703000101.

Wright, Steven. 2002. *Storming Heaven: Class Composition and Struggle in Italian Autonomist Marxism.* London: Pluto Press.

Wyatt, Sally. 2008. "Technological Determinism Is Dead; Long Live Technological Determinism." In *The Handbook of Science and Technology Studies*, edited by Edward J. Hackett, Olga Amsterdamska, Michael Lynch, and Judy Wajcman, third edition, 165– 180. Cambridge, MA: MIT Press.

Wyatt, Sally. 2010. "Challenging the Digital Imperative: Inaugural Lecture." In *Science and Technology Studies at Maastricht University: An Anthology of Inaugural Lectures*, edited by Karin Bijsterveld, 147–174. Maastricht, The Netherlands: Datawyse/Universitaire Per Maastricht. http://www.virtualknowledgestudio.nl/staff/sally-wyatt/inaugural -lecture-28032008.pdf.

Wyver, John. 1995. "Beyond Television: A Vast Network of Pipes." *Convergence: The International Journal of Research into New Media Technologies* 1 (2): 19–22.

Y Combinator. 2012a. "Summer Founders Program [Application Form]." Blog post on the Y Combinator website. http://web.archive.org/web/20050319010616/http://www .archub.org/sfpapp.txt.

Y Combinator. 2012b. "Summer Founders Program [Call for Participation]." Blog post on the Y Combinator website. http://web.archive.org/web/20120319004820 /http://ycombinator.com/old/sfp.html.

Yuill, Simon. 2008. "All Problems of Notation Will Be Solved by the Masses." *Mute: Politics and Culture after the Net,* May 23, 2008. http://www.metamute.org/editorial /articles/all-problems-notation-will-be-solved-masses

Zandbergen, Dorien. 2011. "New Edge: Technology and Spirituality in the San Francisco Bay Area." PhD diss., University of Leiden. https://openaccess.leidenuniv.nl /handle/1887/17671.

Zuboff, Shoshana. 2015. "Big Other: Surveillance Capitalism and the Prospects of an Information Civilization." *Journal of Information Technology* 30 (1). https://doi.org/10 .1057/jit.2015.5.

INTERVIEWS

Adkins, Ian, and Iain Major, November 26, 2009. Founders of Bites from Bytes, the first firm based on selling RepRap derivatives. Clevedon, UK.

Anders, Lambert, September 19, 2011. One of two founders of Techzone. New York, NY.

Bohac, Jiri, September 14, 2008. Contributed mechanical inventions for Ronja, user of Ronja. Prague, Czech Republic.

Bowyer, Adrian, November 24, 2009. Initiator of the RepRap project. Bath, UK.

Budington, William, December 26, 2016. Hacktivist with the Electronic Frontier Foundation. Berlin, Germany.

Darkveggy, March 22, 2014. Member of the *print* hacklab in Dijon. Dijon, France.

de Bruijn, Erik, November 11, 2009. Core developer of RepRap and founder of the firm Ultimaker. Eindhoven, Netherlands.

de Stigter, Johan, September 30, 2008. Running a company selling wireless equipment, sponsor of Ronja. Phone interview.

Dulzet and Joseba, August 16, 2020. Escamot Espiral participants. Calafou, Spain.

ehmry, April 20, 2021. Hacker and free software developer. He emphasized in a later conversation that the quote is not his original invention, but part of internet lore. Indeed, "IRC never dies" is a popular phrase on the internet. LAG lab, Amsterdam, Netherlands.

Elias, Michael, September 10, 2008. Experimented with Ronja design, vendor and user of Ronja. Prague, Czech Republic.

Gullik, Webjörn, August 10, 2008. Experimented with Ronja design. Phone interview.

Hecko, Marcel, December 17, 2008. Developed PCB for Ronja, administrator of a nonprofit wireless network, user of Ronja. Bratislava, Slovakia.

Higgs, Forrest, November 3, 2011. Former core developer of RepRap, initiator of Tommelise 3D printer. Phone interview.

Hitter, Markus, aka Traumflug, September 11, 2011. Maintainer of Gen 7 electronics. Phone interview.

Horky, Jakub, January 17, 2009. Vendor of Ronja. Prague, Czech Republic.

Hudec, Jan, December 8, 2008. Tested the first versions of Ronja. Prague, Czech Republic.

Jones, James, August 11, 2020. Initiator of CubeSpawn, an open-source, flexible manufacturing system. Phone interview.

Jones, Rhy, November 26, 2009. PhD student of Adrian Bowyer at the University of Bath. Developed multiple materials for printing. Bath, UK.

Kamenicky, Tomas, December 4, 2008. Developer of a second generation of free space optics. Prague, Czech Republic.

Kincheloe, Lawrence, November 10, 2009. Promoter of open manufacturing. Phone interview.

Kolovratnik, David, December 14, 2008. User of Ronja. Prague, Czech Republic.

Krishnan, Arun, October 17, 2008. Developer and user of Ronja in Kerala, India. Phone interview.

Kulhavy, Karel, November 16, 2008. Also known as "Clock," initiator and main developer of Ronja. Zurich, Switzerland.

Laufer, Mixael, September 4, 2020. Initiator and spokesperson for the Four Thieves Vinegar Collective. Phone interview.

Leman, Batist, November 12, 2009. Promoter of RepRap in Flanders, Belgium. Initiator of a hackerspace in Ghent, Belgium.

Lunar, May 23, 2013. Member of the *print* hacklab in Dijon, France, Debian developer. Toulouse, France.

Mars, Marcell, April 12, 2014. One of the founders of net.culture club mama in Zagreb, Croatia. Calafou, Spain.

Members of the Autistici/Inventati collective, June 2013. Dublin, Ireland.

Michnik, Jakub, December 17, 2008. Vendor of Ronja. Brno, Czech Republic.

Mockridge, Patrick, August 17, 2020. Member of CubeSpawn team, developer of a virtual control system. Phone interview.

Myslik, Lada, January 9, 2009. Main developer of Crusader. Prague, Czech Republic.

Nemec, David, December 14, 2008. Vendor of Ronja. Chrudim, Czech Republic.

Nipe, Gustav, December 23, 2009. Promoter of RepRap in Sweden. Initiator of the Swedish Pirate Party's "pirate factory." Lund, Sweden.

Olliver, Vik, May 4, 2010, and August 12, 2020, phone interview. Built the first proof of concept of RepRap, among many other things. Auckland, New Zealand.

Palmer, Chris, aka Nophead, March 17, 2010. Made key contributions to the extruder head, among other things. Holds the record in selling RepRap printed parts. Manchester, UK.

Pettis, Bre, September 20, 2011. One of three founders of MakerBot Industries, the second oldest company selling RepRap derivatives. New York, NY.

Polak, Michael, January 16, 2009. Running an internet service provider, user of Ronja and Crusader. Prague, Czech Republic.

Prusa, Josef, September 19, 2011. Principal architect of the Prusa RepRap design. New York, NY.

Ribera-Fumaz, Ramon, September 26, 2014. Human geographer at the Open University of Catalunya. Barcelona, Spain.

Ricola, February 1, 2014. Member of the *print* hacklab in Dijon, France. Calafou, Spain.

Seliger, Petr, September 21, 2008. Developed PCB for Ronja, user of Ronja. Prague, Czech Republic.

Sells, Ed, May 7, 2010. Former PhD student of Adrian Boweyr at the University of Bath. Principal architect of the Mendel RepRap design. Auckland, New Zealand.

Simandl, Petr, October 27, 2008. Administrator of nonprofit wireless network and independent developer of open hardware designs. Prague, Czech Republic.

Snajdrvint, Karel, December 14, 2008. Administrator of a nonprofit wireless network, user of Ronja. Chrudim, Czech Republic.

Sykora, Jakub, November 27, 2008. User of Ronja. Prague, Czech Republic.

Tesar, Ondrej, October 5, 2008. Developed PCB for Ronja, distributed light diodes, user of Ronja. Prague, Czech Republic.

Wattendorf, Nick, and Bruce Wattendorf, September 18, 2011. Promoters of RepRap in the New England area, built the third RepRap Darwin machine in the world. New York, NY.

Wolf, Clifford, August 20, 2001. Member of the Metalab hackerspace in Vienna. Vienna, Austria.

Zajicek, Ondrej, December 14, 2008. Administrator of nonprofit wireless network, user of Ronja. Chrudim, Czech Republic.

EMAILS AND MAILING LISTS

Bonne, K, February 8, 2018. Ronja mailing list. https://web.archive.org/web/201305 09101502/https://lists.pointless.net/pipermail/ronja/

Bowyer, A., September 21, 2012. "Fixing Misinformation with Information." Maker-Bot blog. https://web.archive.org/web/20121026131004/http://www.makerbot.com /blog/2012/09/20/fixing-misinformation-with-information/#comments

Dalton, J., June 30, 2003. "Re: Ronja Page." Ronja mailing list. https://web.archive .org/web/20130509101502/https://lists.pointless.net/pipermail/ronja/

Hitter, M., August 21, 2020. Email communication.

Hoeken, Z., August 6, 2009. "MakerBot is pioneering distributed manufacturing! Get paid to make parts for future MakerBots." MakerBot blog. https://web.archive.org /web/20100301205700/http://blog.makerbot.com/page/17

Kulhavy, K., June 27, 2003. "Re: Ronja page." Ronja mailing list. https://web.archive .org/web/20130509101502/https://lists.pointless.net/pipermail/ronja/

Kulhavy, K., July 20, 2003. "Re: [info@alphawave.cz: Product Annoucement— AlphaWave Crusader TP 10Mbit LED]." Ronja mailing list. https://web.archive.org/web /20130509101502/https://lists.pointless.net/pipermail/ronja/

Kulhavy, K., September 21, 2005. "New economic model for free technology?" Oekonux mailing list. http://www.oekonux.org/list-en/archive/index.html

Kulhavy, K., June 23, 2005. "Ronja donation model." Ronja mailing list. https://web .archive.org/web/20130509101502/https://lists.pointless.net/pipermail/ronja/

Merten, S., September 27, 2005. "New economic model for free technology?" Oekonux mailing list. http://www.oekonux.org/list-en/archive/index.html

Obadal, K., November 1, 2004. "Ronja Business." Ronja mailing list. https://web .archive.org/web/20130509101502/https://lists.pointless.net/pipermail/ronja/

Pettis, B., August 11, 2009. "The User is The Factory." Thingiverse blog. https://web .archive.org/web/20100505002954/http://blog.thingiverse.com/2009/08/page/2/

Prusa, J., February 15, 2011. "Prusa Mendel and the Clonedels." Open 3DP. https:// web.archive.org/web/20150926191531/http://depts.washington.edu/open3dp/2011 /02/prusa-mendel-and-the-clonedels/

Ppeetteerr, March 17, 2009. "Open questions for Zach." Rep Rap forum. https:// reprap.org/forum/read.php?1,22284,22501#msg-22501

RussNelson, August 27, 2007. "Poor economics on front page?" RepRap forum. https://reprap.org/forum/read.php?1,4570,4741#msg-4741

Spacexula, July 21, 2011. "Loaner program." RepRap forum. https://reprap.org/forum /read.php?1,91501,91501#msg-91501

Unknown, November 3, 2003. "AlphaWave přináší Ronju." CZFree mailing list. https://www.czfree.net/forum/showthread.php?s=&threadid=3064&perpage=16 &highlight=&pagenumber=1

INDEX